From Playgrounds to PlayStation

From Playgrounds
to PlayStation

The Interaction of Technology and Play

Carroll Pursell

Johns Hopkins University Press
Baltimore

© 2015 Johns Hopkins University Press
All rights reserved. Published 2015
Printed in the United States of America on acid-free paper
9 8 7 6 5 4 3 2 1

Johns Hopkins University Press
2715 North Charles Street
Baltimore, Maryland 21218-4363
www.press.jhu.edu

Library of Congress Cataloging-in-Publication Data
Pursell, Carroll W.
      From playgrounds to PlayStation : the interaction of technology and
play / Carroll Pursell.
         pages cm
      Includes bibliographical references and index.
      ISBN 978-1-4214-1650-2 (pbk. : alk. paper) — ISBN 978-1-4214-1651-9
(electronic) — ISBN 1-4214-1650-6 (pbk. : alk. paper) — ISBN 1-4214-
1651-4 (electronic) 1. Games and technology—History. 2. Toys—History.
3. Electronic games—History. 4. Play—History. 5. Recreation—History.
I. Title.
      GV1201.34.P87 2015
      790—dc23   2014031358

A catalog record for this book is available from the British Library.

Special discounts are available for bulk purchases of this book.
For more information, please contact Special Sales at 410-516-6936
or specialsales@press.jhu.edu.

Johns Hopkins University Press uses environmentally friendly book materials,
including recycled text paper that is composed of at least 30 percent post-
consumer waste, whenever possible.

For James and Mary

If I've bored my readers, may they forgive me, since I myself have been hugely amused.   LOUIS SEBASTIEN MERCIER

Contents

Acknowledgments

Some of the material in this book has previously appeared in print. Parts of chapter 1 are taken from my essay "Toys, Technology and Sex Roles in America, 1920–1940," *Dynamos and Virgins Revisited: Women and Technological Change in History*, ed. Martha More Trescott (Metuchen: Scarecrow Press, 1979), pp. 252–267. Chapter 2 appeared in an earlier form as "The safe and rational children's playground: Strategies and technologies since the nineteenth century," *History Australia*, 8 (December 2011), 47–74. Most of chapter 3 appeared in a previous version as "Fun Factories: Inventing American Amusement Parks," *ICON: Journal of the International Committee for the History of Technology*, 19 (2013), 1–19. Some paragraphs from chapter 4 were taken from my essay "The Long Summer of Boy Engineering," *Possible Dreams: Enthusiasm for Technology in America*, ed. John L. Wright (Dearborn: Henry Ford Museum & Greenfield Village, 1992), pp. 34–43. Some of the material in chapter 7 was taken from "The Art and Commerce of Video Games," *Technik zwischen artes und arts, Festschrift für Hans-Joachim Braun*, ed. Reinhold Bauer, James William, and Wolfhard Weber (Munster: Waxmann, 2008), pp. 149–157.

I want to express my deepest gratitude to my editor, Bob Brugger, who was in on the beginning of this project many years ago and whose encouragement and patience have never flagged. I also acknowledge my profound indebtedness to my partner, Angela Woollacott, whose love and example have been indispensable.

From Playgrounds to PlayStation

Playing with Technology

This book is about the ways in which technology and play interact. Play can, of course, take place without any technologies whatever, but a good deal of play involves a range of technologies and, importantly for some people in some circumstances, technological creativity is itself a kind of play. My focus here is on play in the United States, and mainly in the nineteenth and twentieth centuries, though some European precedents are also outlined. Both technology and play are transnational phenomena, with influences and artifacts moving more or less freely between different parts of the world, and I have felt free to refer to many of these.

While not all play involves technology (though arguably all technology can be played with), as a common and familiar activity it offers a particularly useful lens through which to observe how technology was changing during the two centuries covered. This text discusses inventors and manufacturers, entrepreneurs and consumers, ideologies and evolving social customs. As a fundamental aspect of life itself, play has been deeply embedded in the American experience. Gary Cross, a leading historian of play, has isolated three themes in the story of play in America: the ways in which changes in work and its timing have shaped play; the ways in which play has been shaped by changing technologies and commercialization; and, finally, the way in which "play and its meanings changed along with childhood and the family."[1] All three of these basic themes are critical to the stories I tell in the following pages.

The existing serious literature on this subject consists mainly of academic books on the psychology of play (almost exclusively concerning children), on the sociology and history of toys, on sites of entertainment such as Coney Island and Disneyland, and on the sociology of leisure activities, as for example do-it-yourself home improvement and hot rod modification. There is also a large scholarly literature on issues of childhood, with academic journals focusing exclusively on both children and play. No attempt is made in this volume to summarize this vast literature, but it serves as a background for my own focus on the technology of play.

This text is basically a social history, but since such issues as creativity, professional development, childhood and gender, race and class are intimately involved, a measure of cultural analysis has been undertaken to clarify some of these larger issues. Neither play nor technology, as categories, can be precisely defined, since the two are historically both contingent and contested. On the topic of baseball, for example, I take technology to include not only balls, bats, and mitts, but also the rules, the playing diamond, and the stadium that surrounds it. Even more broadly, the technology of baseball, as it is enjoyed in the United States, would include the airplanes that transport the teams and the television that broadcasts it to a wide audience far from the field of play. On the other hand, it has been argued that, as a sport, with rigid rules, schedules, and commercial constraints, baseball is no longer play at all but much more like work—perhaps even for the children who are marshaled into ubiquitous and mandatory Little League games and practices. The relationship between work and play is critical to this subject and explored in various chapters.

As with any book, this narrative is the result of a range of choices. Some particular technologies, for example, the bicycle, fit nicely into more than one chapter: in this case it's considered as a toy, a hobby, a sport (including an extreme sport), and even at one time a daredevil loop-the-loop ride in amusement parks. Here I have chosen to discuss it within the sports category, but another author could easily have made another, and just as legitimate, choice.

Finally, out of the myriad of toys, hobbies, sports, amusement park rides, and so forth, I have had to choose a relatively few to discuss in any detail. Again, my choices may not be the same as the ones that another person would have made, but I have tried to provide a mix of the most important, the most familiar, and, I confess, sometimes simply my own favorites. I have also chosen to write about a fair number of them (some might say too many) because I wanted to give an impression of the richness (and even excitement) of the range of devices we have had at our disposal and for our pleasure.

This series of related chapters begins by covering toys, paying special attention to their gendered presentation. Some boys play with dolls and some girls with Erector sets, but such gender violations are often viewed as atypical if not subversive. And what is seen as "natural" for each sex has itself changed. At one time, baby dolls were said to appeal to girls' natural desire to nurture and playing with them reinforced that trait. In more recent times, Barbie dolls are seen to appeal to girls' natural desire to shop for clothes. And G.I. Joe dolls (or action figures) are seen as appropriate for boys. In the first chapter, the discussion ranges from dolls to BB guns, paying particular attention to scaled-down home appliances, wagons, scooters and similar transportation toys, and to sets or kits, such as for construction (Lincoln Logs) or science (especially chemistry).

The next chapter provides a survey of, with some thoughts on, the American playground reform movement, which began in the late nineteenth century and remains with us. Drawing heavily upon both the period's concern for child welfare and the intellectual imperatives of Frederick Winslow Taylor's Scientific Management movement, urban neighborhood play environments were engineered to provide wholesome and efficient activities for children of working-class families. Over time, the emphasis on efficient play was challenged by a concern for aesthetic design and creative opportunity for children. Ironically, either way this attempt to provide a safe environment for play itself came to be seen as dangerous.

A third chapter covers commercial pleasure parks, from Coney Island to Disneyland. As John Kasson famously has shown, such sites are great collections of machines which often imitate the technology of transportation and production with which their working-class patrons are already all too familiar. More recently, Arwen Mohun has shown the fascinating ways in which "rides" were carefully engineered to commodify risk while providing the reassurance of safety. Fairylands for children are also destinations for having fun through rides, shows, and fanciful environments.

Next, our attention turns to those leisure activities, especially crafts, that arose in the nineteenth century to bridge the widening gulf between home and work spaces. Again gender rules prescribed appropriate activities, with heavy tools such as saws and hammers being reserved for men, and light tools, like sewing or knitting needles and paint brushes, reserved for women. Of course, girls and boys were encouraged to learn their appropriate skills and spend free time engaging in appropriate activities. In the twentieth century, new technologies presented opportunities for new hobbies—operating ham radios, building model airplanes, assembling hi-fi equipment, modifying cars, and, by the end of the century, cooking gourmet meals using a wide range of expensive kitchen utensils.

Inevitably we must look to the rise of intercity professional sports, especially baseball and basketball, and to intercollegiate sports such as football, all of which developed in late nineteenth-century America, and to such popular pastimes as golf and tennis. The country's elaborate railroad network, along with the telegraph and the rise of mass-circulation newspapers and magazines, made national sports possible, and a cadre of pioneer sporting goods manufacturers helped to create and to supply the demand for the equipment required. The rationalization of the terms of competition, and the standardization of playing fields and equipment, were critical to the success of this cultural revolution. Technological changes, such as "high-performance" swimsuits and "anchored" putters, continue to challenge the concept of the level playing field.

We think of traditional sports as demanding skill and stamina, but what of activities that use new devices to defy gravity or stress human endurance? One has to wonder at the late twentieth-century rise of "Xtreme Sports," from the older surfboard to skateboards, roller blades, high-tech bicycles, hang gliders, windsurfing boards, snowmobiles, snowboards, and parkour. These are all, of course, hypermasculine adaptations of technologies more commonly used by both sexes but now given new context—edgy danger. Ironically, although these activities are surrounded by an aura of countercultural youth alienation, they came to prominence through the influence of corporate marketing efforts. In a nice postmodern gesture, the sport of Extreme Ironing comments ironically on the entire enterprise.

Finally, the nexus of play and technology invites discussion of electronic games, from Pong to PlayStation. Recently, a heralded "Tot-Com" boom has revealed the size and importance of electronic toys for kids. Since everyone seems to agree that children's play is formative, and therefore controversial, violent action games and failed "girl" alternatives provide a clear window to the problematical nature of play and technology. At the same time, the creation of games has provided new venues for both art and music. Electronic games, it seems, have reproduced many of the joys and the alarm of those older toys discussed in the first chapter.

Toys for Girls and Boys

With an outfit of this kind, you are doing something real—something every boy wants to do.   —GILBERT CATALOG *Boy Engineering* (1920)

[Toy appliances] enable the small girl to exactly counterpart her mother's industries.   —CHRISTINE FREDERICK (1928)

P lay, it has been said, is children's work, and toys are their tools. In all places and ages children have played with things, some found by children, some fabricated by them, and some provided by parents or other adults. Today these might include a just-emptied rolled-oats carton salvaged from the kitchen, a knocked together wooden wagon set on cast-off baby buggy wheels, or a gaudy heavy plastic gym set of Chinese manufacture. In a child's imagination, the category of "toys" is expansive indeed. But just as much adult work has historically been considered to be gender-specific, so has play, and therefore toys, been socially limited as suitable for only one or the other of the sexes. In theory, and often in practice, boys' and girls' toys have been so identified and even kept segregated in toy stores: there are the girl aisles and the boy aisles, with the latter being more numerous and showing a greater range and variety of items.

In real life, however, there are often circumstances in which the distinction is not maintained. In 1893, Frances Willard, the American feminist and long-time head of the Woman's Christian Temperance Union, was not in good health. A friend gave her a bicycle, and she was determined to teach herself to ride. In a short essay describing that experience, published two years later, she chose to begin with an account of her childhood a half century before.

With her hair cut short, Frank, as she insisted upon being called, had what she termed a "romping" childhood, and spent as much time as possible outdoors. She wrote that she "very early learned to use a carpenter's kit and a Gardener's tools, and followed in my mimic way the occupations of the poulterer and the farmer, working my little field with a wooden plow of my own making, and felling saplings with an ax rigged up from the old iron of the wagon-shop." At sixteen, however, she was brought indoors and dressed properly: she wrote in her journal, "Altogether, I recognize that my occupation is gone." She recovered that joy of playing only with the gift of a bicycle late in life, and the realization that, in learning to ride it, as she said: "I had made myself master of the most remarkable, ingenious, and inspiring motor ever yet devised upon this planet."[1]

No toy, of course, so defined girls' play, both before and after Willard's time, as did the ubiquitous doll. Although parents, and even children themselves, could, and often did, make serviceable dolls by hand, most came from Europe and particularly Germany.[2] Some women in the nineteenth century turned to doll making as a business and, rejecting both the elaborate and sometimes fragile (with heads made of porcelain, for example) European dolls and the more technologically ambitious dolls of male manufacturers, concentrated on dolls that were often soft and made by female artisans working in an almost preindustrial manner. More important, their more intimate knowledge of girls and their play gave them an additional advantage. Another guide to design came not from the preferences of the girls themselves but from new notions of "scientific motherhood" current in the Progressive period. Girls, however, whether from the middle- or working class, had their own ideas of play, and dolls became objects as much to contest as to inculcate ideologies of girlhood and regimes of play.

It was the male inventors and manufacturers who seized upon the modern industrial materials, techniques, and attitudes to shape the world of dolls. This branch of the doll industry looked remarkably like any other, with inventors competing to create new products. R. C. Purvis patented his in 1901, for example, claiming that "my invention relates to an improved construction of a doll, which is preferably made out of sheet metal, whereby it is rendered more durable and improved in various details." His doll also embodied a mechanism that held the eyes (which opened and closed) in position while also supporting the teeth, which gave the doll an "attractive and life-like appearance."[3]

A similar set of values was evident in a 1904 advertisement in the journal *Playthings* by the Metal Doll Co. of Pleasantville, New Jersey.[4] Along with a photograph of the factory, it featured the "JOINTED 'ALL STEEL' DOLL," which was claimed to be "indestructible" and to have removable wigs and either fixed and movable eyes. Most notably, the company advertised that the dolls were "Jointed at Neck, Shoulder, Elbow, Wrist, Thigh, Knee and Ankle." A quite contemporary

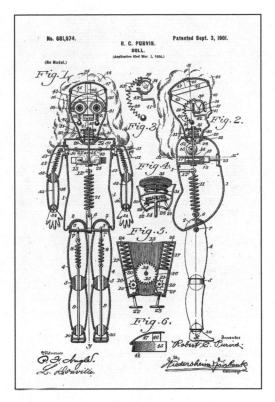

In the late nineteenth century, doll making,
which had been dominated by women,
attracted increasing numbers of male inventors
who mechanized the toy in an attempt to make
it more "lifelike." Drawing of Patent No. 681,974
for a sheet-metal doll, issued to Robert C.
Purvis of Laurel Springs, NJ, September 3, 1901.

movement was built into D. Zaiden's "Mechanical Doll" patented in 1921. It was, he claimed, "a novel construction of mechanically operated doll adapted to perform movements in simulation of the so-called 'shimmy' dance." Specifically, its arms and shoulders were made to shake, presumably in a joyful and provocative manner.[5]

The flexibility of the Jointed All-Steel Doll was in striking contrast to the hugely popular and ultra-feminine Barbie doll, which was introduced in the United States in 1959, but remarkably similar to the male Action Man that appeared in 1964 under the name G.I. Joe. From the beginning, Action Man had

In addition to designing dolls that were more mechanical, the male manufacturers newly drawn to the business were actively industrializing what was still in part a hand process. This 1890 illustration shows a female labor force at work on Thomas Edison's "Talking Dolls," which used a recording playback device of his invention. Created in 1877 and put on the market in 1890, the dolls were a commercial failure. *Scientific American*, 62 (April 26, 1890), 257.

twenty ball-and-socket joints, which meant that he could be posed to appear to be performing a wide variety of strenuous activities. Barbie, on the other hand, had joints only where the arms, legs, and head were attached, severely limiting her actions to little more than posing. Over the years she was given more movement, starting in 1965 when her knees were made bendable so that she could do weight-loss exercises. This feature was later dropped, but others were added as her lifestyle changed through time.[6] Altogether, dolls, from rag through steel to plastic, remained one of the most popular and played with of toys.

A host of new toys became available in America in the early decades of the twentieth century, linked to the marketing of a large number of consumer durable goods which, in conjunction with contemporary innovations in consumer credit, advertising, and marketing techniques, transformed the way in which most people lived. From the automobile, to the radio, to the electric iron, mass production was matched by mass consumption of new technical marvels.

With surprising rapidity, the new adult technologies were scaled down for children. Toy vehicles, tools, appliances, and construction sets quickly introduced children to the marvels of owning and using modern technology. Not surprisingly, the gender ideology of the adult world—which dictated which devices were to be used by which sex—was equally pervasive in the child's world of play.

Toys depicting modern science and technology have been ubiquitous at least since the nineteenth century. Even a casual glance at the toys available for children reveals two striking facts: first, that children are encouraged to follow the scientific and technological fads of their elders; and second, that these are often advertised as being more appropriate for one sex than the other. The Gilbert Company, for example, had an ecology kit on the market during the heyday of the environmental movement in 1973, just as it had had a wireless telegraph outfit in 1930. Also in the mid-twentieth century, Tonka was marketing a kind of anti-ecology toy: a model snowmobile. Tonka also had its own recreational vehicle, the "Mighty-Tonka Winnebago." Advertisements showed a little boy apparently "driving" the vehicle to a campsite, while a little girl took care of the family once they had arrived. One Tonka innovation from the same period seemed strangely anachronistic. During the same year that Charlie Chaplin's classic film *Modern Times* was re-released, and *Newsweek* reported that "Boredom on the Job" drove 500 of the 4,000 workers at one automobile assembly plant to heroin addiction, Tonka unveiled its "Assembly Line Kit." The company boasted that each of the cars "can be assembled, played with, taken apart and reassembled. Again and again."[7]

These examples point to several important facts about toys. First, in all ages and places, toys have been used not only to amuse and entertain, but also "as socializing mechanisms, as educational devices, and as scaled down versions of the realities of the larger adult-dominated social world."[8] In the United States, studies have indicated that "by the age of three or four, boys and girls show decided preferences for appropriately sex-typed activities, toys, and objects." It has been pointed out further that among those cultural artifacts of our society that help to form and strengthen patterns of children's play—the media, formal education, direct parental instruction—"none is so constant and concrete in its impact upon children's play as children's toys."[9] This is, of course, the way gender works: girls and boys are instructed early in what is "appropriate" for each, and not surprisingly children quickly internalize the appropriate preferences.

A look at American toys marketed during the decades between the world wars leads to the conclusion that this use of toys to socialize children into what were considered appropriate gender roles was equally prevalent then. This is especially obvious when one looks at those toys that were thought to embody the principles of modern science and technology. During these two decades, contemporary observers commented on three basic changes that influenced the world of toys. First was a dramatic change in technology itself, especially in the familiar terms of both capital and consumer goods. The airplane, automobile, gasoline tractor, radio, and a host of electrical appliances became common adjuncts of modern life. Second, it was widely asserted that formal education was becoming not only more widespread but more practical as well. Learning by doing and the democracy of experience were hailed as ushering in a new generation of bright, pragmatic, flexible, and innovative young Americans.[10]

And third, there was a burst of growth within the American toy industry itself. Before the Great War, handmade German toys had been prominent on the American market, and toy sales had been largely seasonal, concentrating on the Christmas trade. By the end of the twenties, a highly mechanized, aggressively merchandised American toy industry had grown up to challenge and largely displace the German product, and the selling season had been somewhat smoothed out across the calendar.

In terms of the new technology, girls' toys are most easily described because they were simpler, fewer in number, and concentrated in the areas of cooking, cleaning, and other branches of housewifery. Such toys were not new. A toy kitchen, now in the Metropolitan Museum of Art, was probably made in New York in the late eighteenth century. The Museum of the City of New York contains a toy stove, utensils, and an African American cook, the presence of which in 1884 reinforced racial as well as gender stereotypes.

During the 1920s, many household tasks were electrified: between 1924 and 1930, the number of users of electrical household appliances more than doubled.[11] Mother's new appliances were quickly scaled down for daughter, so that she too could get used to the joys of living electrically. An article in the December 1928 issue of *American Home* featured "Grown-up Accessories for Small People," and referred to little girls as either "young housewives" or "small housekeepers."[12] According to the author, a leading advocate of Taylorite efficiency in the home,

> Mothers will welcome a new and interesting development in toys to gladden
> the heart of the little girl. Boys always have been liberally supplied with outfits
> and playthings which moved and worked, and which they could use construc-
> tively. Such practical gifts have brought boys much fun, because in playing
> with them they could imitate the many admired activities of Father and other

grown-ups. But until recently the small girl has been forced to remain satisfied with tiny dishes, pots, and pans, and with toys of Lilliputian size which she could only pretend were "just like Mother's."

The "young housewife," pictured sitting in her rather isolated (but color-coordinated) kitchen, had a "small scale range [which] is as perfect in detail and operation as that in her mother's own kitchen. It may be plugged safely into any outlet, and is guaranteed to bake, brew, and brown everything from a cake for her father's birthday to a fairy feast for the doll's party." The various tools and appliances, we are told, "bring happiness because they enable the small girl to exactly counterpart her mother's industries." Such toys, in the words of the article, "satisfy the little girl's love of home activities." Since "all little girls love to arrange and rearrange furniture and room furnishings," these too were available, especially "the popular 'dresser' which every true woman adores, no matter her age be six or sixty." By 1929, the famous New York toy store, FAO Schwarz, offered an electric range, along with the more traditional sewing machine.

Hasbro introduced its Easy-Bake Oven for little girls in 1963, using a 100-watt light bulb to provide the heat for baking cupcakes and other items. The technology remained essentially unchanged for fifty years, until the ban on incandescent bulbs necessitated a new heat source. In 2011, the company debuted its new Easy-Bake Ultimate Oven, with a heating element much like that in a real oven. The vice president for global brand strategy and marketing for Hasbro Girl Brands stated, "it is important to us that the new oven looked more like a real appliance than just a toy, so we incorporated a style similar to what you see elsewhere in the kitchen." Oddly, they then "gave it a great color—purple—and some cool graphics" as well. It was, they said, "one of those toys that has proven its longevity and popularity, and it's a good investment in play."[13]

Even though the range of toys designed for them was small, girls did on occasion appear in advertisements for boys' toys, but in such cases the girl was seldom active. In one typical advertisement from 1933, a girl, clutching her appropriate dolly, is admiring the skill and boldness of her brother, who is playing with a Stanlo set, one of the many construction toys produced for boys beginning in the World War I era.[14]

Science and technology toys for boys were much more varied than the range of appliances for girls, and can be divided into four rough categories: tools, vehicles, construction sets, and science outfits. Carpentry tools remained the standard items in that field, representing a craft still resistant to change. In 1920, the A. C. Gilbert Company offered thirteen such sets, ranging from the No. 701 Gilbert Carpenter's for Boys, at $3.50, to the No. 760 Special Tool Chest. This last item, "built especially for the Government to be used during the great war," sold for $50. "Extremely well built," according to the catalog, "it makes a chest that any boy can well be proud of."[15]

Closely tied to the gift of tools was their use, and around Christmas time the idea of making toys was popular. A 1934 article in *Parents Magazine* offered advice on "Gifts for Them to Make." The author, incidentally, was credited with having written two books: *Handicraft for Girls* and *The Boy Builder*.[16] A movement to allow boys to take home economics at school, while girls could take shop, was endemic during these years, and in 1932 the vice principal of a high school in South San Francisco reported on one toy-making experience there.[17] The shop teacher had gotten his boys into making toys, but, as the vice principal noted, "eighteen years of tradition had practically 'sealed the fate' of girls as far as the industrial arts courses were concerned. But there was a single girl who had nerve enough to talk the matter over with the principal. . . . The fact must not be overlooked," he added, "that this girl was a 'martyr' to the cause, for she certainly received—well, as high school jargon has it—she received the 'razz.' Only for a few weeks, however, for because of her earnestness, her nonchalant attitude, and the 'walking Mutt and Jeff' which she made, she soon became the object of envy for many girls." The next semester, fifteen girls enrolled in shop—but were taught in a segregated class.

The category of vehicles was varied and in some ways very traditional. Gilbert applied for patents for three toy vehicles in 1919. One was really just a novel way of connecting the axel to the body of a vehicle similar to a tricycle without pedals.[18] A second was for a scooter, the patent drawing for which shows rows of holes in the various sections, though the description does not mention that it would be put together from a kit.[19] The third was for a "Toy Vehicle," designed to come as a kit of metal parts similar, as he said, to common "construction toys . . . usually sold in the form of a building set." It was essentially an Erector set, but with parts "so organized and constructed , that full size playthings for children, and more particularly vehicles of various types adapted to support or carry children, can be readily built up in various combinations." The patent drawing shows a wheelbarrow, again with holes and bolt fasteners.[20]

A classic American wagon, the Red Radio Flyer, by its very name captured some of the drama and excitement of post–World War I consumer technologies. Antonio Pasin began production of wooden wagons in Chicago in 1917, calling them Liberty Coasters in honor of the Statue of Liberty which had welcomed him to the country. In 1927, he switched to steel bodies for his wagons and adopted techniques of mass production from the automobile industry, with 1,500 bright red wagons coming off the assembly line daily during the Great Depression. Eventually production expanded to include tricycles, scooters, bicycles, and other forms of juvenile transportation. The red wagons remained the major product, however, with occasional new models being offered, such as the streamlined Zephyr and, in the 1990s, a "Quad Shock Wagon" suggesting the popular sports utility vehicles of the time.[21]

Although wagons, sleds, scooters, and bicycles were still popular, small cars, such as those made by the Wolverine Co. in 1924, represented "the latest in juvenile automobile design." As the firm declared, "we conceived the idea that the most outstanding mode of transportation in a modern child's mind was an automobile."[22] Although such vehicles might appear to be relatively gender-neutral, wagons and such usually were shown with a girl being pushed or pulled about, or, in some cases, with a girl subordinated by perspective. The Auto-Wheel Coaster of 1921 was the subject of a concerted sales campaign aimed at boys. As the catalog predicted, "if Tom has an 'Auto-Wheel,' you may be sure that Dick and Harry want one too!" Auto-Wheel Clubs were said to have enrolled 25,000 boys, all of whom received copies of the *Auto-Wheel Spokes-man*, "a lively little publication full of good live *stuff* that every boy likes to read." In addition, reflecting the rise of credit buying among adult consumers, the clubs "help boys to buy their 'cars' and often actually advance club funds for this purpose."[23] On a smaller scale, in 1920 Gilbert offered a toy tractor, stating that "every wide-awake boy knows what wonders the Tractor has accomplished, and what a tremendous aid it is in the great farming districts of the West. You boys want to see how these up-to-the minute machines work."[24]

Science kits, sets, or outfits were also popular with boys during the twenties and thirties. The 1920 catalog of the A. C. Gilbert Company, entitled *Gilbert Boy Engineer*, listed numerous kits, such as that for civil engineering. "With an outfit of this kind," Gilbert wrote, "you are doing something real—something every boy wants to do." Other kits promoted hydraulic and pneumatic engineering, magnetism, sound, meteorology, machine design, signals, and electricity. The kit for light experiments perhaps was closer to magic (the field in which Gilbert got his start) than modern science. Boys were told: "You simply press a button or turn a switch and you have light. Do you know why—or where it comes from? No! Because it's electricity."[25]

Wheeled vehicles like wagons and tricycles proved consistently popular. This advertisement is unusual for showing both a boy and a girl enjoying the exhilaration of speed, though not surprisingly the boy is featured in the foreground. Cover illustration for the 1921 catalog of Auto-Wheel Coaster Co., *Illustrating Auto-Wheel Coasters and Auto-Wheel Convertible Roadsters and Fleetwing Steering Sleds* (North Tonawanda, NY, 1921).

The chemistry set probably was the most popular kit. Such sets had been on the market for many years: a "Boy's Own Laboratory" was offered to the public in 1882; it included 54 chemicals and 30 pieces of apparatus "for performing endless experiments in Chemical Magic," was warranted "free from danger," and sold for six dollars.[26] World War I has been called a "chemists' war," and the 1920s saw a tremendous growth in the American chemical industry. As it related to children, the Porter Chemical Company made the message explicit in its 1928 Chemcraft catalog—today boys play at chemistry, tomorrow men hold scientific posts in industry. "Today," it explained, "no matter what profession a man follows, he is greatly handicapped without a knowledge of chemistry. The manufacturer, the farmer, the tradesman, the professional man, the scientist, all have constant need of chemical knowledge. In the home the housewife who knows nothing of the chemistry of the food which she prepares or the materials which she uses daily is seriously handicapped."[27]

It was not all serious learning with no fun, however. Despite a traditional view of science which celebrated the progress from "the Alchemist of Old to the Modern Chemist," chemical magic was often emphasized. "Chemistry," it was claimed, "is also a spectacular science and many chemical phenomena are most startling and mystifying to the layman." Facing what must have been a perennial parental worry, the firm admitted that "chemistry is sometimes looked upon as a dangerous profession, but this is not the case. Contrary to an old popular idea, a chemical experiment does not necessarily result in an explosion."[28] Once again, there seems to be no good reason why chemistry sets should have been limited to boys, but advertisements typically featured only boys.

Such science toys were claimed to have a profound impact, at least upon some children. The head of one toy company was said to believe that "very often the careers of great men in various scientific fields have found their first inspiration . . . in the playthings which amused and fascinated them in childhood." George Ellery Hale, the American astronomer, was singled out as a contemporary scientific luminary who was benignly influenced by toys.[29] One study, published in 1922, attempted to discover the influence of after-school science clubs, as the nearest controllable approximation to free play with science kits. After claiming that Newton and Faraday had "developed their love, interest, yes, and their fundamental background and experience in play with science toys," the author concluded that, among other benefits, "extra-curricular activities in science represent a type of purposeful activity which encourages originality and inventiveness and habituates boys to the experimental procedure."[30]

Among all categories of toys, those dealing with construction were said to be the most American. In actual fact, the English Meccano sets invented by Frank Hornby apparently antedated similar American sets. According to an oft-repeated story, Hornby observed a crane working outside his train window while on a Christmas trip, and proceeded to develop toy components that could be assembled into cranes—or a variety of other machines and structures. His obituary noted that his imagination was "kindled to the possibility of mechanical toys which would appeal to the boy mind in general," and that this toy "could be put together by any boy of average intelligence." The notice also recalled that he began his works in "one room and he had only one girl [presumably also of at least average intelligence] to assemble the parts."[31]

By the time he died in 1936, Hornby was publishing his Meccano instructions in seventeen languages, and, as early as 1915, his annual competition for new models was attracting 10,000 entries. The winner that year was a toy loom sent in from New York City. "Meccano," said the company, "does teach boys engineering. All the time they are building models they are acquiring knowledge which may some day prove of the greatest practical value to them."[32]

In his *Autobiography*, Frank Lloyd Wright told of his mother's visit to the Centennial Exhibition held in Philadelphia in 1876. She "made a discovery. She was eager about it now. Could hardly wait to go to Boston as soon as she got home— to Milton Bradley's."[33] Bradley operated a color lithography shop in Springfield, Massachusetts, and in 1860 began to manufacture colorful board games. By the 1870s he had become the first American manufacturer to produce croquet sets, including wickets, mallets, balls, stakes, "and an authoritative set of rules to play by that Bradley himself had created from oral tradition and his own sense of fair play." By that time, Bradley had also become deeply involved in the Kindergarten Movement as advocated by the German Friedrich Froebel. He produced the colorful bits of paper that Froebel thought children should use to make creative shapes. He also manufactured the brightly colored wooden shapes that Froebel called "Gifts" and also recommended children play with.[34] Wright's mother saw to it that her son had a full set of "Gifts," including maple building blocks.

Several American construction toys from this period have remained popular. The Chicago architect John Lloyd Wright, son of Frank Lloyd Wright, introduced Lincoln Logs for building frontier forts, cabins, and other structures. The son, like the father, had been raised on the principles of Friedrich Froebel. Besides being influenced by the popular enthusiasm for the frontier and the memory of

Abraham Lincoln (himself born in a log cabin), John Wright also claimed that he was inspired by his father's use, in designing the Imperial Hotel in Tokyo, of a new technique called floating cantilever construction.[35] His 1920 patent for "Toy-Cabin Construction," he claimed, was an improvement in "toys and more particularly educational toys calculated to develop a child's constructive inclination." It took "the form of a multitude of members or parts in the form of logs, so shaped and treated as to provide material for the building of diminutive log cabins and other like structures."[36] The Lincoln Log firm also held competitions for new designs, and Wright asserted that "a real American boy with a keen brain is just about the smartest and most original thing alive. Precedent and custom mean nothing to him. He is bold and courageous in the execution of his ideas for this reason." It was also his belief that "the future salvation of the American toy industry lies mainly in the manufacture of construction toys, these being the kind the foreign makers know the least about."[37] In addition, because the construction kits contained large numbers of identical parts, they could be more easily mass-produced by machines, reducing costs and the need for skilled labor.

Another construction toy made in large quantities by specialized machines, introduced in 1914, was the Tinker Toy. The president of Toy Tinkers, Inc., in 1924 claimed that in their factory, "design, plans and preparations, automatic machinery, expert labor, simplification, standardization, synchronizing of production, all work together to give us a manufacturing advantage that cheap labor, cheap materials, and generations of toy-making experience cannot offset." Tinker Toys, he boasted, had had twenty-nine imitators, none of whom had been able to operate at a profit.[38]

The leading American construction toy perhaps, and the one most like Meccano, was the Erector set. The toy was the invention of the remarkable A. C. Gilbert, who was once described as blending "the familiar qualities of Frank Merriwell, Theodore Roosevelt, Peter Pan and Horatio Alger." Frail as a child, Gilbert followed the familiar regimen of strenuosity, became a star athlete at Yale, and set a world's record in the pole vault in 1908. The next year he received a medical degree, although he never practiced. Instead, he began to manufacture apparatus for magic shows—a childhood passion. Throughout his life he appears to have been obsessed with the virtues of manliness and the need for competition, qualities often associated with science and especially engineering. "Everything in life is a game," he once reflected, "but the important thing is to win." Like Hornby, he received his inspiration for the Erector set (1914) while looking out a train window at girders being erected. Years after making patterns for the first Erector set, Gilbert remarked that "I've remained a boy at heart and only introduced items that appealed to me. I figured they would appeal to all boys."[39]

The introduction to his 1920 toy catalog offers a glimpse into his way of thinking. "I feel," he wrote, "that every boy should be trained for leadership. It is only the bright-eyed, red-blooded boy who has learned things, done things, dared things beyond the reach of most boys who will find the way open to really big achievements. . . . My toys are toys for the live-wire boy, who likes lots of fun and at the same time wants to do some of the big engineering things—things that are real—things that are genuine."[40]

An eventual link between traditional construction toys, like Erector sets and Lincoln Logs, and computer games was provided by the plastic bricks made by Lego (which means "play well" in Danish, and "I put together" in Latin). Ole Kirk Christiansen opened his own carpenter shop in Billund, Denmark in 1932. He began making step ladders, ironing boards, and similar products, along with wooden toys made from scrap lumber. Gradually the toys took over, and in 1934 he renamed his company Lego, since wooden blocks were his main product. In 1944, his factory burned and when he rebuilt it after the war, he invested in a plastic injection molding machine, the first in Denmark, in 1947. Two years later, the plastic Lego Automatic Building Bricks were first introduced, produced with either four or eight studs and in four different colors. There is some evidence that Christiansen's inspiration for Lego bricks came from England. There the child psychologist Hilary Harry Fisher Page took out British patents in 1940 and 1949 for self-locking building bricks, sold under the name Kiddicraft.[41]

One of Christiansen's accomplishments was to create a successful plastic toy in a period when many preferred wood and "plastic" sometimes carried seriously pejorative associations. Another was to invest in expensive injection molding equipment, and finally to improve the design of the bricks. In 1958 he developed a new "stud-and-tube" coupling system which improved the stability of Lego structures. Another innovation was to integrate the twenty-eight different sets Lego had on offer, and the eight vehicles along with various add-on elements, into a "System of Play" which enabled all of them to be used interchangeably in new combinations. Starting with only two sizes of bricks, by 2004 Lego was manufacturing 12,400 different pieces.

By the turn of the twenty-first century there was a diminishing interest among children in traditional "construction" toys like Erector sets and Lincoln Logs. With its large variety of pieces, Lego led the transition of construction to narrative and role playing. If an Erector set had been used to build a ferris wheel, for example, that might have been the end of the play. With Lego, it was only the beginning, since children could then integrate buildings, vehicles, and figures into story lines to continue the play.[42] Indeed, the Lego company made sets to construct boards, which then provided a platform on which board games could be played. Later Lego entered into agreements with other companies such as Walt Disney to produce video games, and opened theme parks in Denmark, England, Germany, Florida, and California.[43] Of course, Lego could not control all the uses to which their iconic bricks were put. An artist created large figures from the bricks and toured an entire show of the pieces. More problematically, a book was published, *Forbidden LEGO*, which showed how to make real weapons, such as catapults, from the toy.[44]

Most surprisingly, the company launched Mindstorms 2.0 in 1998 and Mind-storms NXT in 2006. Each box contained 619 elements with which one could build one's own robot: a microcomputer, various sensors, servo-motors, a "CD with easy-to-use software with icon-based programming language," and other necessary items. The kit could be used to build four different robots, including "a humanoid robot, easy to assemble and with multiple functions; it walks and turns, dances, talks, can see and avoid obstacles, can grasp and distinguish be-tween different colored objects. *Robogator* is an animal robot that moves like an alligator. It will protect its area and jumps forward and snaps at anything that comes too near. Watch out!"[45] Almost immediately, enthusiasts emerged who hailed the kit as "a long-overdue merging of construction toy and computer," and they began to modify and extend Lego's program.[46]

Lego set off in a quite different direction in 2011, when it launched a special line of products to try to capture a piece of what was called "the girls' toy market worth $267 million last year." Lego Friends remained a construction toy, which "mirrored the boys' experience," but was designed to also appeal to girls' "interest in remodeling and redesign, and themes like community and friendship." Build-ing sets featured five friends, each with a different personality and set of interests, living in a town called Heartlake City.[47]

In 1938, Gilbert bought out the American Flyer electric train line and added it to his offerings, thereby providing a centerpiece for the sometimes elaborately built environments that some hobbyists produced long before Lego sets were available. After World War II, Gilbert introduced an Atomic Energy Labora-tory. The Erector set remained the firm's most famous product, however. Like Meccano, the Gilbert Company used a wide variety of devices to reach the boy market. During the Christmas season of 1933, for example, Gilbert's "big illus-trated 'Look-Em-Over' Book" contained an entry blank for the Erector model contest: second prize was a new Chevrolet and first prize, a trip to "the Panama Canal, or Boulder Dam or the Empire State Building or any other engineering project in the United States." Boys were also urged to tune in to the Sunday eve-ning radio program, "'Engineering Thrills' True stories about real engineers and their hair-raising adventures in digging the Panama Canal, building bridges and skyscrapers."[48]

Considering that guns were a ubiquitous feature of American life, it is not sur-prising that toy weapons were early on and widely available. One of the most pop-ular was the cap pistol. These metal guns have a trigger and hammer, which drops on a small package of explosive powder to make a realistic pop when the trigger is pulled. One popular type, at least from the 1930s, was designed like a six-shooter and often sold with a belt and holster. In the 1940s, cowboy movie stars often lent their names to the guns, and movie theaters, which featured "Westerns" and catered to the children who watched them, sometimes required young customers to check their guns at the box office when tickets were purchased. In the 1950s, television programs augmented this trend, and Western stars such as Roy Rogers, Gene Autry, Hopalong Cassidy and Davy Crockett allowed their names to be as-sociated with specific pistols and rifles.[49] When Westerns on television and in the movies waned in popularity, so too did cap pistols.

Another very popular toy weapon was the BB gun, especially those made by the Daisy Manufacturing Company. Unlike cap pistols, which could (sometimes dangerously) look like real weapons but only made a noise, air rifles like the Daisy actually shot a small metal ball and thereby blurred the distinction between toys and "real" technologies. Indeed, in 1970 Daisy introduced pneumatic "high-powered adult air guns," because "young adults were demanding more sophisti-cated equipment."[50]

Daisy had its origins in the Plymouth Iron Windmill Company, of Plymouth, Michigan, which had begun production in 1882. The company's general manager had become enthusiastic about a gun made by the Markham Air Rifle Company, also of Plymouth, allegedly exclaiming, "Boy, that's a Daisy!" Since windmill sales were disappointing, the company decided to give an air rifle away as a premium for each windmill sold. The company began to make the guns themselves, and the offer proved so successful that it was turning out 50,000 guns a year by 1890. Five years later it got out of the windmill business and changed its name to the Daisy Manufacturing Company.[51] In 1911, Charles F. Lefever, who had invented a pump action air gun, joined the company. The pump gun was first manufactured in 1914 and proved enormously successful, with 8 million sold before it was discontinued in 1979.[52]

Like cap pistols, Daisy BB guns flourished because of their associations with popular cultural figures, real and imagined. In the 1930s, the cowboy circus star Buzz Barton and the film star Buck Jones both lent their names to popular Daisy guns. Then, in 1933, the Buck Rogers Rocket Pistol appeared, followed the next year by the Buck Rogers Disintegrator Pistol. Both were based on the popular comic strip hero. Perhaps its most iconic BB gun, the Roy Rogers model, appeared in 1940, associated with another popular cowboy hero. It proved so successful that a million were sold in 1949.[53]

The line between air rifles used as toys and those used as weapons was further blurred in 1963 when Daisy, along with the U.S. Jaycees, launched a shooters education program for children. More than ten million boys and girls were enrolled in the first years, and it was later expanded through cooperation with the Boy Scouts, American Legion, National Rifle Association, 4-H clubs, high school Junior Reserve Officers Training Corps, and other groups.[54]

Always alert to what boys might like, by 1920 Gilbert had begun producing war toys. Six months before the end of World War I, he applied for a patent for a toy submarine, receiving it in 1920. It was designed to be propelled by a spring-driven screw propeller, and could fire torpedoes as well. "If it is desirable to make my toy submarine submerge, all that is necessary is to tilt the diving rudders," he pointed out.[55] In 1918, Gilbert applied for a patent for what he called simply a "Toy Gun" but described as "a toy machine gun designed to amuse and entertain boys and to train them in the military use of machine guns."[56] The catalog's description boisterously proclaimed: "if there ever was a real live-wire toy for the red-blooded boys, this is it. Say, you can have more genuine sport with this machine gun than anything I know of. It's the real thing."[57] What he called his "most significant" war toy was also one of the few he marketed for girls—"a nurse's outfit, with cap, arm band, bandages, adhesive tape, scissors, and a bottle of soda-mint tablets."[58]

"War toys" were controversial at least as early as World War I. A collection of short stories by the English author Saki (Hector Munro), published in 1923, was titled "The Toys of Peace" but was dedicated to the 22d Royal Fusiliers. The title story told of a mother who did not want her two boys playing with model soldiers, weapons, warships, and so forth, so she asked her brother to bring them "peace toys" for Easter presents. Though dubious of the experiment, the uncle arrived with a box full of the sorts of "little toys and models that have special bearing on civilian life in its more peaceful aspects": a municipal dust bin, a model of John Stuart Mill, another of the Manchester branch of the Young Women's Christian Association, a set of hop-poles, a small ventilator for sewers, and similar items. The boys, who had hoped for a Somali camel corps and model Albanian soldiers ("they have got jolly uniforms"), were visibly disappointed. When told that these were toys, meant to be played with, the boys responded, "But how?" The uncle could only suggest that they might make a game with Mill, and the other figures, contesting a seat in Parliament. Peeking at the children at play a short time later, he discovered that they had converted the toys and models to weapons and troops, and were creating a narrative of war. "The soldiers rush in and avenge his death," cried one of the boys excitedly, "with the utmost savagery. A hundred girls are killed." Ironically, Saki, who was keen to see combat, was killed by a sniper on the Western Front in November 1916.[59]

Saki's experiment was declared a failure, but the issue returned. With the escalation of the war in Vietnam in the mid-1960s, the slogan "No War Toys," emblazoned on posters, bumper stickers, and lapel buttons, became commonplace. Twenty-one-year-old Richard Register dropped out of Occidental College in Southern California and founded the organization, No War Toys, in 1965. He met Bob Dylan, received help from Joan Baez, and got The Doors to play at a fund-raiser two days before they opened at a club on Sunset Strip. His motivation was simple and straightforward: "I thought," he later wrote, "a lot more was going on than most people recognized when parents gave little boys war toys and smiled while the kids pretended to kill one another."[60]

From model airplanes to chemistry sets, from Tinker Toys to Erector sets, it was confidently asserted during the interwar years that such toys helped children, specifically boys, become prepared for life in the world of modern science and technology. "Our children," wrote the president of American Flyer in 1921, "are growing up to manhood and womanhood to face an intense industrial era—a machine governed world—which will call to its aid the highest efficiency of science to achieve mass production at minimum costs." American toys, he concluded, instruct the child "in the fundamentals of the great mechanical forces with which he must cope in his adult days."[61] In his autobiography, A. C. Gilbert quoted a Yale professor who claimed that "one of the biggest factors in the growth of the chemical industry in the United States had been Gilbert Chemistry sets," and cited "the hundreds of letters" he had received "from engineers" who told him that "their first interest in their profession started with an Erector set."[62]

This education for the modern world of science and technology was, of course, different for the two sexes. Household toys, often replicas of the latest electric appliances, were directed at girls, while science and construction toys, tools, and vehicles were aimed at boys. The distinction was obvious and almost without exception. The influence of this ubiquitous training in "appropriate" gender roles would be difficult to measure but must have been enormous. The constant urging of boys to investigate, experiment, innovate, and be red-blooded, live wires was designed to bring each boy to his highest level of capability. Girls, on the other hand, were not so encouraged, but were given toys (such as dolls) and appliances that prepared them for a lifetime of cooking, washing, and cleaning.

An investigation of "Children's Toys and Socialization to Sex Roles" uncovered a large area of agreement between the assumptions of toy manufacturers (the producers and advertisers), parents (and purchasers), and children (the ultimate consumers), concerning male and female sex roles and the toys appropriate for each.[63] The study concludes that "boys' toys are viewed as the most active, the most social, and relatively high in terms of complexity. Girls' toys, on the other hand were not seen as the 'most' anything. They were the least complex (most simple), the least active (most passive), and were virtually tied with the boys' toys in their relatively low ratings on creativity and education." Some of the more striking data showed that three out of four chemistry sets pictured only boys on the boxes, and the remained pictured both boys and girls— none showed girls only. On craft sets, there was a correlation between the sex of children on the boxes and the "high" or "low" technology involved—bead craft showed girls, electronics showed boys. Boys received more gifts in absolute numbers, their gifts cost more, and were more widely varied, according to this study.

Toy stores routinely make shopping easier by grouping "boys' toys" in one area and "girls' toys" in another, frequently in separate aisles. It was not until 2011 that Hamleys, the iconic London toy store, ended its practice of putting the girls' toys on the third floor (and marking it with pink signage) and showing the boys' toys on the fifth floor (with blue signage), following a campaign by the neuroscientist Laura Nelson.[64] Three years later, however, much remained unchanged. One observer noted that the first floor had a display of dress-up costumes, with a complete bridal outfit (including a corsage), a pink cowgirl outfit, a pink waitress uniform, and a pink and purple superhero costume, for the girls. For boys there was a blue police officer uniform and a blue superhero costume.[65] One floor up one could find, conveniently grouped, pink housecleaning sets, cooking utensils, and hairstyling kits. The Barbie display featured "construction sets for a ballet studio, a fashion boutique, an ice-cream cart and a luxury mansion." This stood beside the display for Walt Disney Princess products. In 2011, there were some 26,000 different Disney Princess products on the market: the world's largest franchise for girls aged two to six.[66]

Still, it was reported in 2013 that the company that had the Toys "R" Us franchise in Nordic countries was reforming the layout of its Stockholm store to eliminate the gender-specific displays. One newspaper revealed that "play kitchens stand opposite train sets; baby strollers are piled beside a stack of toy guns; My Little Pony stares at swords and ninja costumes. . . . Even the Barbie house—that last redoubt of the candy-pink—is under pressure from a Lego display." The company explained that "we did many interviews with children about toys, and for example they complained they couldn't get a sword for their Barbie, or a baby stroller for Spiderman . . . Why not leave it up to the child's fantasy?"[67]

Regardless of the theory of play, there has been, through time, a consistent belief that play was somehow closely connected to the later life of children—a particular application of the old adage that "as the twig is bent . . . " The lives of engineers appeared to hold a special attraction for commentators. The argument went in both directions. Adults connected to toys, such as the inventor-manufacturer A. C. Gilbert, were convinced that when children played with, for example, his Erector sets, they were not only learning engineering skills and attitudes but were also more likely to choose engineering as a career. In apparent support of this proposition, engineers were often cited as claiming that their career decision indeed had been decisively influenced by early play with Erector sets, model airplanes, or similar scaled down technology. A closely related argument also moved in both directions: Gilbert insisted, for example, that in playing with these toys, children were actually not just learning but *doing* engineering; in other words, doing real work. At the same time there was often a suggestion that engineers, when at work, were often "playing" with ideas and materials, enjoying their tinkering and puzzle-solving. The point has been made that a "playful" frame of mind is an important aid to creative design and innovation for engineers, as well as for scientists and architects.

The engineer Henry Petroski has reported that a survey of engineers who went on to become "American business leaders" revealed that many of them "recall with fondness playing as children with chemistry sets and construction toys such as Lincoln Logs, Tinker Toys and Erector sets." Additionally, many reported that as children they had been interested in "how things work," often taking apart their toys and then reassembling them. Petroski admitted that he too had played

This 1933 advertisement for Erector sets also gave information on Gilbert's radio program "Engineering Thrills," his "Look em-over" book, and details about the 1,021 prizes available for new machines built with Erector parts. The copy carried his assertion that "I am positive every red-blooded boy will want to be an Erector Engineer this year." *Parent's Magazine*, 8 (December 1933), 2.

with such sets, as well as toy trains. He seems to have had a particular fondness for his bicycle, taking it apart to see how it worked but also making repairs and doing general maintenance. "Little did I know it then," he later wrote, "but tinkering with bicycles was preparing me for a career in engineering as surely as were the mathematics, science and mechanical-drawing courses I was taking in high school."[68] Based partly on the survey, partly on the testimony of older engineers who, when receiving awards and honors, recall their play as children, and partly on his own experiences, Petroski declared, "I believe a good number of engineers of my generation [he was born in 1942] and older did play with Erector sets, electric trains and bicycles, not to mention chemistry sets, radios, electronics kits, Heath hits and the like."[69] Frank Lloyd Wright wrote of playing with building blocks as a child: "the smooth shapely maple blocks with which to build, the sense of which never leaves the fingers: *form* becoming *feeling*."[70]

Petroski's evidence is anecdotal, but ubiquitous. An obituary for Stanley Hiller,

Jr., the helicopter innovator, is a case in point. The notice remarks that, as a child, "his playground was his father's workshop . . . ," and that he "soon started to make and fly model airplanes." He enrolled in the engineering college of the University of California at Berkeley, but left after one year to pursue his interest in helicopters.[71] The Nobel physicist Martin L. Perl complained in later life that his parents would not buy him an Erector set, but he was able to play every Saturday with a cousin who had one. "He also," said Perl, "had electric trains. I loved to build with the Erector set, I loved to build toys and models out of wood, I loved to draw mechanical devices, even those I could not build. I loved to read the magazines, *Popular Mechanics* and *Popular Science*. I loved all things mechanical; cars, trucks, derricks, trains, and steam ships. I was in love with mechanics, and I still am." He said that his parents would not buy him a chemistry set, though he was interested in chemistry. "Strangely," he concludes, "for a person who became a physicist, I was not interested in amateur radio or in building radios."[72]

For the 1999 National Engineers Week, the Silicon Valley/San Jose *Business Journal* asked, "Engineers, how did you play as a child?" One replied that "I took bikes apart to see how they worked. Then I put them back together." Another said that he had first had a Teddy bear, but after that, "Tinker Toys and the erector set were my best toys." A third replied that "I had two hobbies growing up in Israel: radio and electronics." An atypical answer was that "my childhood was not filled with erector sets. I did not build things. My father did that. I played with Barbies." She went on, however, to take a degree in engineering. Other responses included: "I took everything apart and put it back together when I was a kid to see how it worked. I also loved to play 'civil engineer,' damming up water as it ran along the street"; "I had a chemistry set and an erector set"; "building things. That was it. In addition to Tinker Toys and a erector set, I built a tree house. . . . In high school I bought a Heath Kit Audio System."[73]

A report from the FERMI Lab that same year followed the traditional path: when the staff was asked what they wanted for Christmas as children, "Erector sets, the ones with the brass nuts and bolts, the red tool kits, and the electric motors, head the list. Chemistry sets follow close behind." One reported building model airplanes, then "promptly blowing them up with the help of the faithful chemistry set." Building crystal radios and playing with trains were both popular. One significant report came from "a theorist, not an experimentalist." We are told that "although his parents equipped him with the traditional Erector set, he found he lacked the patience for tightening all the little bolts and nuts. Legos gave him the means to build fantastic constructions without the tedium of bolting them together."[74]

This shift from Erector sets to Legos was not much reported, but was sometimes controversial. In 2001, Sir Harry Kroto of Britain, who was awarded the Nobel Prize in chemistry, deplored the change. Speaking of the front room of his house when he was a child, he recalled that

> it filled up with junk and in particular a Meccano set with which I "played" endlessly. Meccano . . . is called Erector Set in the US. New toys (mainly Lego) have led to the extinction of Meccano and this has been a major disaster as far as the education of our young engineers and scientists is concerned. Lego is a technically trivial plaything and kids love it partly because it is so simple and partly because it is seductively coloured. However it is only a toy, whereas Meccano is a real engineering kit and it teaches one skill which I consider to be the most important that anyone can acquire: This is the sensitive touch needed to thread a nut on a bolt and tighten them with a screwdriver and spanner just enough that they stay locked, but not too tightly that the thread is stripped or they cannot be unscrewed. On those occasions (usually during a party at your house) when the handbasin tap is closed so tightly that you cannot turn it back on, you know that the last person to use the washroom never had a Meccano set.[75]

The arguments of both Petroski and Kroto are tinged with nostalgia and a not uncommon belief among an older generation that the younger has somehow not measured up to the high mark they set. Nevertheless, the notion of play, in both engineering education and practice, has gained some attention. In 1992, the engineer and historian Eugene Ferguson published a book, *Engineering and the Mind's Eye*. "The mind's eye," he wrote, "the locus of our images of remembered reality and imagined contrivance, is an organ of incredible capacity and subtlety. . . . [It is] the organ in which a lifetime of sensory information—visual, tactile, muscular, visceral, aural, olfactory, and gustatory—is stored, interconnected, and interrelated. We get to know things through a series of sensual interactions."[76]

Ferguson believed that workers—"machinists, millwrights, carpenters, welders, tinsmiths, electricians, riggers, and all the rest—supply all made things with a crucial component that the engineer can never fully specify. Their work involves the laying on of knowing hands. It is sad that engineering schools teach contempt, not admiration, for those hands."[77] Ironically, he noted, in the 1950s, just when engineering schools were abandoning courses aimed at giving students hands-on experience in designing and building machines, they were also gripped by a "creativity craze" which sought somehow to teach that elusive concept. Ferguson believed that more reliance on the mind's eye was needed, not the heavily hyped CAD (computer-aided design) "coming with ready-made software that would promise to solve whole classes of problems with a minimum of knowledge on the part of the operator."[78]

A sense of what had been lost is given in a 1923 article in *Popular Mechanics*. "Toy Models Help to Solve Many Problems for the Engineer and Inventor," it announced, adding that "Miniature Ships, Airplanes, Motors, Houses, and Building Blocks Are Found of Assistance in Visualizing Their Ideas." At first "intended only to amuse children, toys, or miniatures, are now being used everywhere by engineers, architects, builders, and inventors to visualize their problems and to quickly test out their ideas." The demand for these models was said to have spawned "a new industry that employs thousands of men throughout the country." A part of this market was provided by schools which, "realizing that the eye is many times quicker than the ear, . . . have turned to the use of what would have been considered mere playthings just a few years ago, in instructing both boys and girls."[79]

Nearly a century later, and half that long since the craze for creativity, schools that trained scientists, mathematicians, and engineers were being encouraged to implement "playful methods of learning." One advocate explained that "the pedagogical value of play does not lie in its use as a way to teach children a specific set of skills through structured activities called 'play.' Rather, play is valuable for children primarily because it is a *medium* for development and learning." Because by the early twenty-first century "playful methods of learning have almost disappeared from school classrooms, and active creative, extended playtimes during recess, at home, and in neighborhoods have also diminished," it was imperative that it be reintroduced at all levels. "When adults play with ideas," she continued, "use creative techniques, accomplish risky or unusual feats, employ models to exemplify their mental worlds, and allow themselves to be truly comfortable and creative in their environment, they are also using the medium of play, although they may call it something else." Playful learning in schools, she suggests, should become "the wave of the future."[80]

Not surprisingly, "playful learning" became harnessed to computer technologies. Advocates of IvT, which combines "paper-based, physical processes with new, virtual digital design environments that complement traditional approaches," claim that it creates a powerful "'think, play, do' schema." The architect Frank

Gehry is presented as a role model for such a process, because he "plays with ideas when selecting the optimum shape for his buildings." Gehry, they claim, "trained using drawings and physical models. He is a craftsman who loves to work with his hands, sculpting shapes in a range of materials such as cardboard, clay, and plasticine." He was initially uncomfortable with computers, but his designs proved "difficult to engineer and build." Traditional CAD systems were inadequate to his needs, but in 1990 Gehry procured a more powerful software, the CATIA system used by Boeing for airplane design, which he ran on an IBM RISC 6000 computer. It made possible, they say, "the overlapping of thinking, playing, and doing, but it also demanded a closer degree of cooperation between architect, contractors, suppliers and subcontractors." The process "demonstrates how to take ideas from the mind's eye and prepare them for action and implementation."[81] It is a cooperation of which Eugene Ferguson might have approved.

Since in the United States engineering is one of the most sexually segregated professions, this entire narrative of early influences has focused on boys and their play. Despite considerable effort during the 1970s to recruit more women into engineering programs in colleges, universities, and technical institutes, the rate of female participation has remained low. In 2013, women made up only 14 percent of the profession, and there was evidence that this already low figure had been falling during the first decade of the new century.[82]

Debbie Sterling, who graduated from Stanford University with a degree in Mechanical Engineering/Product Design, has taken seriously the argument that construction toys help steer children into engineering careers. In 2013 she began to market GoldieBlox, a construction toy she designed "from a female perspective" with which she hope to "disrupt the pink aisle and inspire the future generation of female engineers." She believed that "there are a million girls out there who are engineers. They just might not know it yet."[83] In a series of kits, a girl in overalls, named GoldieBlox, builds various machines with the help of animal friends. The toys include a storybook, character figures, design templates, and various wheels, axles, blocks, washers, a crank, and a ribbon. The fact is that most of the boys over several generations who played with Erector sets did not become engineers. The same will no doubt be true of girls who play with GoldieBlox, but that is not to say that the experience will have no effects later in life.

Chapter Two

The Safe and Rational Playground

Every child has the right . . . to engage in play and recreational activities.
— U.N. CONVENTION, RIGHTS OF CHILD, ARTICLE 31 (1989)

All play is a voluntary act. Play to order is no longer play.
— J. HUIZINGA (1944)

Recognizably modern playgrounds had been attached to some German schools during the nineteenth century, and one was built in Manchester, England, as early as 1859, but during the years around the turn of the twentieth century, according to Allen Guttmann, "the wholesale intervention of adults utterly transformed the informal world of traditional children's play."[1] The campaign for playgrounds in the United States, which began as part of the Progressive Movement in the late nineteenth century, though backed by a range of supporters, early fell into the hands of efficiency-minded people who tried to apply the principles of Scientific Management to children's play and its dedicated venues. After World War I efficiency faded, and into the 1930s, playgrounds that encouraged children's creativity and sense of design took hold. Throughout the twentieth century, concerns about safety, especially as it pertained to the technology of playground equipment, remained a persistent theme, leading to what one critic referred to as an undermining of childhood itself.

The playground movement that arose in the United States at the turn of the twentieth century was fueled in large part by a perception that such new venues would be safer than the urban streets where so many children, especially of emigrant groups and the working class, spent their play time.[2] Though many games, such as hide-and-seek and tag, needed no equipment, a range of devices from swings and slides to seesaws and sandboxes quickly became expected features of the new playgrounds. Despite the subsequent rise of professional playground supervisors and an attempt to make playgrounds into schools of scientific play and nurseries of efficiency, by the end of the century they were considered dangerous, full of technological traps and mechanical hazards.

Urban streets were the first playgrounds for the thousands of children, especially of emigrant groups and the working class, who lived in the teeming industrial cities of the nation.[3] Playmates abounded; as the comedian George Burns recalled, the building where he lived housed sixteen families, and each family had eight to ten children. Apartments in the tenements were cramped and crowded, siblings shared not only rooms but beds, and there was simply nowhere to play indoors. Adding to the problem, there was a chance that adult females in the family might be using the space to do piecework while another family member, or possibly a paying boarder, might work a night shift and need to sleep during the day. Spending the time between school and dinner playing outside was the logical, necessary, and enjoyable solution. Following the homosocial mores of the adults, girls claimed the stoops as their own territory, where they could sew, talk, and keep an eye on younger siblings. Older boys controlled the middle of the streets for their games and activities, while younger boys took the spaces between the two.

Children who wanted to play baseball in the streets had to create baseball diamonds out of found objects (manhole covers, fire hydrants, lampposts) and this need extended to other forms of play. Historian David Nasaw has described how "children had to scrounge their equipment and then some. Garbage pail lids were made into sleds, bicycle wheels into hoops, discarded cans and used bags into footballs; baby carriages were transformed into pushcarts and wagons, and scraps of wood too insignificant to be used for kindling at home were burned in bonfires in discarded lunch pails."[4]

Johan Huizinga has written that "All play moves and has its being within a playground marked off beforehand either materially or ideally, deliberately or as a matter of course. . . . All [playgrounds] are temporary worlds within the ordinary world, dedicated to the performance of an act apart."[5] This marking off of streets by children was a direct challenge to the adults who claimed them as their own world of commerce, transport, and social intercourse. Peter and Iona Opie, chroniclers of children's play, noted that the young and adults had been sparring over public spaces at least since the Middle Ages, and that children marked their claimed space by "screaming, scribbling on the pavements, smashing milk bottles, banging on doors, and getting in people's way."[6] Playing by their own rules implied breaking those of adults.

Adult activities not only were often irksome to children trying to play, they could be dangerous as well. The role of the streets as transportation corridors created a tangle of wagons, horses, streetcars, buggies, bicycles, and, eventually, cars and trucks. Darting out in front of any one of these might prove fatal. The potential danger was compounded when children, either to save money or for the thrill of it, jumped on and off passing vehicles like streetcars while they were moving, or just rode on the outside, clinging to part of it. More than one child was killed trying to steal bits of ice off the iceman's delivery wagon. The danger posed by passing vehicles was part of the reason that, at the turn of the twentieth century, reformers began to demand that children be cleared from the streets and given purpose-made playgrounds.

In 1910, the *New York Times* carried an article titled "Ask Mayor to Clear Streets of Children." Colonel E. S. Cornell, identified as secretary of the National Highways Protective Association, charged that "last month there were thirteen children killed in the streets here [in New York City]—stricken down, simply because they had no place to go except the street, where they could get a breath of fresh air." Cornell and his association advocated that the city allow children to play in the many vacant lots and abandoned schoolyards that dotted the city.[7] In fact, many New York schools had no yards at all, but at least a few had experimented with putting playgrounds on their roofs. One playground, operating as early as 1898, was said to have "sand piles, parallel bars, horizontal bars, hitch and kick standards, see-saws, swinging ladders, overhead ladders, and basketball."[8] Seward Park had begun as an area of rubble from slums demolished in 1898. At the behest of the Outdoor Recreation League, the site was eventually cleaned up and dedicated as a playground in 1903.[9]

Cities across the United States were already beginning to make some provisions for playgrounds, especially in dense working-class neighborhoods. Settlement houses were established by philanthropists especially to aid immigrant working families, beginning in the 1880s, and by the turn of the century there were hundreds. Often they operated such social services as day-care centers, well-baby clinics, and playgrounds. These last were often merely empty dirt lots, lent out by philanthropists. Municipal governments had since the mid-nineteenth century established large urban parks, such as Central Park in New York City and Golden Gate Park in San Francisco. The latter, established in 1870, opened a dedicated playground in 1887. Refurbished in 2007, it retained its famous and popular long, dual cement slide built into a hillside. Smaller parks, usually less than four acres in area, were also established, either for the peaceful contemplation of nature, or as active playgrounds.

Boston was an acknowledged leader in the movement, with the Massachusetts Emergency and Hygienic Association operating playgrounds in schoolyards during the summer months. These typically contained "sand gardens" and swings, along with picture books, small blackboards, small toys, and games for younger children. What was called "the good example of Boston" was followed by Providence, Rhode Island, Baltimore, where the United Women of Maryland maintained several playgrounds in schoolyards, and Philadelphia. There the City Parks Association opened a playground in 1894, and by 1898, the city board of education itself was operating twenty-seven playgrounds in the afternoon, under the supervision of the school janitors.[10]

In 1893, the West End District of the Associated Charities of Chicago cooperated with Jane Addams's famous Hull House settlement to operate "a large playground in an empty lot, equipped with swings, see-saws, giant stride, and sand bins." In 1896, "under the auspices of the university settlement of the Northwestern University, a large and splendidly equipped playground was opened, which will accommodate 3,000 or 4,000 children. . . . Numerous swings, large and small, giant stride, see-saws, sand piles, etc., afford ample amusement for the children, who fairly swarm there." A police lieutenant from the local precinct claimed that "not less than fifteen lives have been saved from the electric car since the establishment of the playground, and juvenile arrests have decreased fully 33 1/3 per cent."[11]

Chicago's Municipal Science Club began studying the need for additional, often smaller, parks in 1898, and the following year the mayor established a Special Park Commission. In 1904, Chicago began to create hybrid spaces with landscaping for adults, playgrounds for children, and often playing fields for organized or "pick-up" sports. The celebrated landscape firm of Olmsted Brothers and the architectural firm of D. H. Burnham & Co. were hired to design facilities that often had running tracks, wading pools, sand courts, and field houses, in addition to playground equipment. In 1905, the Davis Square Park, near the Union Stock Yard, opened a field house, a gymnasium for women, another for men, as well as meeting rooms, a library, and a cafeteria. It attracted five million visitors in one year.[12] Such elaborate facilities, of course, defined the upper end of playground design.

Playgrounds in Chicago's schoolyards were each supplied with "a sand pit, wood paving blocks for building, swings, see-saws, and parallel bars. In addition some had ladders, climbing ropes, ring-toss, and other games and apparatus." A special effort was made to "interest the children and educate them in rational play." One observer noted that many children seemed "ignorant" of play and showed "a lack of initiative" when confronted with this bounty of equipment. "The swings," he noticed, "were the most popular form of amusement, doubtless because the most individualistic, and at first there was little consideration of others." "Work," he concluded, "is a cardinal principle of American life. To this must be added play."[13] This linkage of work with play made it obvious that pieces of playground equipment were a part of the American child's toolbox.

In 1904, the city of Los Angeles appointed a commission to oversee its recently acquired playgrounds. The object was to keep new facilities out of the hands of the people who ran the city's parks. "The purpose of the playground," the *Los Angeles Times* pointed out, "is different from a park in that it is intended to be more than a mere park."[14] The first playground in the city was opened in June of the following year, and contained a house on the grounds for the manager, the man who was "to superintend the games and play of the whole playground." Half the playground was "portioned off for the girls," and contained sandboxes, six "box" swings and five rope swings, a Maypole, seesaws, basketball courts, and "a pretty summer house with tables just the right size and that is actually to be devoted to dolls and their accessories." The boys' portion of the grounds contained an open-air gymnasium with a trapeze, a horizontal bar, parallel bars, a leather horse, a "leather buck," a rope ladder, two punching bags, six traveling rings, peg poles ("to develop grip"), slanting ladders, vertical ladders, and horizontal ladders. There were also handball courts, seesaws, and swings.[15]

Forcing children into "rational play" of course ran against the grain of play on the streets, where kids subverted adult rules and were adept at making up their own. One historian has noted that "while neighborhood kids enjoyed using the equipment provided, they did not relish the adult 'supervision' that came along with it."[16] In part this was because the children did not share the philanthropists' understanding of the city as a place that was dangerous to life, limb, and morals; rather, they found the streets, dumps, railroad yards, and harbor fronts to be places full of potential for adventure and even a bit of profit. But it was also because those same philanthropists had a vision of shaping savage immigrant, working-class children into docile workers, trained to curb their initiative and follow often mind-numbing rules and regulations in worksites increasingly organized around the principles of Frederick Winslow Taylor's principles of Scientific Management.

An attempt by the city of New York to get its charter revised by the legislature so that it could organize playground administration and hire supervisors inspired Lurana W. Sheldon to write a poem, "Recreation by Rule," for publication in the *New York Times*.[17] Noting that the city wished to "Teach the children how to play by scientific force," she continued:

They wish the parks and playgrounds
to be under supervision,
So all the games of children may be sub-
ject to revision.
And as for Park Instructors, Playground
Mentors and the like . . .
They are going to teach the youngsters
how to play with common sense,
And they'll surely make them wretched at
considerable expense.

This wish to regularize play found its signature expression in the attempt, during the Progressive Era, to make playgrounds conform to guidelines akin to the widely admired principles of Scientific Management.[18] Two signal markers of the impending modernity of the nineteenth century were the related trends toward specialization and professionalization. Inevitably, playground advocates decided to form a nationwide organization to foster their goals. The National Playground Association of America was founded in Washington, DC, in April 1906.[19] President Theodore Roosevelt, an ardent advocate of both strenuous activity and national efficiency, was made vice president of the association, while the position of president went to the prominent physical education leader, Luther Halsey Gulick.[20]

Gulick had begun his study of physical education at Oberlin College in 1884.[21] The following year he attended the Sergeant School of Physical Education in Cambridge, Massachusetts, and in 1889 earned his MD from the City University of New York. In 1891, as head of the gymnasium department of the Young Men's Christian Education's Springfield Training School, he assigned one of his students, James Naismith, the task of designing a game around a set of simple rules that he provided. The resulting sport of basketball quickly grew in popularity and complexity.

In 1906, Gulick was the head of physical education for the New York City school system, and served as president of both the American Physical Education Association and the Public School Training Association. The following year he became the chairman of the Playground Extension Committee of the Russell Sage Foundation, a philanthropic organization that was a hotbed of Taylorite ideals of social efficiency. Gulick fit in well, writing a book *The Efficient Life* in 1907 and having already worked to establish standards of athletic accomplishment by taking moving pictures of athletes, in much the same way that Frank and Lillian Gilbreth studied adult workers.[22] Meanwhile, the Russell Sage Foundation was sponsoring books that studied schools as "factories of learning," producing, for example, flow charts to maximize movement through the school's plant.

Working with Gulick was Lee F. Hanmer, who moved from being a physical training supervisor for the New York City schools to field secretary of the new Playground Association of America, and then secretary of the playground extension committee and associate director of the department of child hygiene at the Russell Sage Foundation. In 1912, he was made director of the Foundation's recreation department, a post he held until his retirement in 1937.[23]

Seeking funds to continue its work, the Playground Association turned to the Russell Sage Foundation, which gave it $20,000 and undertook to pay Gulick's salary directly.[24] Almost immediately the executive committee of the Association called for two studies: one of "the relative time that should be given to sleep, to concentrated effort, that is to work, and to play," and a second study that would identify the "plays most beneficial for special groups." The point of this last was to "develop precise timetables so that playground engineers could manufacture and target play programs" at specific groups, such as emigrant and working-class children.[25] In 1913, A. E. Winship, the editor of the *Journal of Education*, published an article in the Playground Association's magazine *Playground*, titled "Science and Art in Play."[26] It was the science he seemed to be most interested in, declaring that "play must be as definite as mathematics. Play must be as scientific as the laboratory. There must be all the relationship of cause and effect that there is in business." Behind science lay the issue of class: another contributor to *Playground* asserted that "there are types of play which can be made the best method of teaching children to work, and that the desire to play . . . can be directed to inculcate habits of industry leading to good citizenship."[27]

Lee Hanmer himself developed and published certain athletic standards and exercises, which, if precisely followed, would result in their accomplishment. Children were to compete, not against each other, but against these standards, and be rewarded with a badge for their success. The process was not unlike that used by the recently formed Boy Scouts of America. These standards were praised in a *Playground* article in 1914: "the aim of the badge test of the PAA is to secure greater efficiency in the lives of growing boys." The author claimed as well that "the present tests correspond to the methods of scientific management . . . they are based on modern principles of efficiency." John H. Chase, who was head social worker at the Goodrich settlement house in Cleveland, Ohio, a center of Taylorist play, told the readers of *Playground* in 1909: "We want a play factory; we want it to run at top speed on schedule time, with the best machinery, and with skilled operatives. We want to turn out the maximum product of happiness, to utilize all the space, to be awake to new inventions, to use our minds for planning and our hearts for enthusing."[28]

The class basis of playground development was clearly evident when historian Roy Rosenzweig analyzed the patterns of working-class leisure in one industrial city, Worcester, Massachusetts. As early as 1879, there were complaints that parks accessible to the working class of the city were neglected, while resources were poured into upper-class neighborhoods. Indeed, while the latter were parks oriented toward the horticultural production of calming and contemplative places of beauty and serenity, workers demanded not parks but playgrounds—venues for active recreation and games.[29] Indeed, these desired "playgrounds" were apparently to be for adult workers as well as their children.

Rosenzweig asserts that in the 1890s the dominant protestant business class was losing its "old Calvinist suspicion of play" and was beginning to embrace "competitive sports and outdoor recreation." In 1917, Worcester's old Parks Commission was reconstituted as the new Parks and Recreation Commission and attention was turned to the control of working-class behavior, especially among children. This urge, coupled with a new academic interest in play, according to Rosenzweig, "made play reform a central project of progressive reformers."[30] In 1908, the Massachusetts Playground Act was passed to ensure that the state had at least one playground for every 20,000 residents. The Worcester Parks Commission "began constructing special play facilities for park users, such as baseball fields, tennis courts, wading pools, outdoor gymnasiums, picnic groves, swings, sandboxes, and seesaws." The commission had previously installed "primitive playground equipment in [one park] . . . in 1898 but by 1909 it was experimenting with the latest steel gymnastic apparatus."[31]

Such facilities, of course, tended to dictate what types of play would occur within them. This structuring of acceptable play was made explicit with the move to hire playground supervisors to direct activities. The Worcester Playground Association, formed in 1910, took as its main mission not the adding of playgrounds or equipment, but their supervision. Paid staff oversaw organized athletics, drama, folk dancing, storytelling, and such crafts as gardening, sewing, and basketry. Playgrounds tended to be dominated by one ethnic group or another, rather than randomly integrated, but efforts at "Americanization" proceeded in all of them, by scheduling American games, singing patriotic American songs, and staging American folk dances. In 1912, it was reported that many children, disdainful of efforts to teach them how to play, either ignored the supervisors or simply stopped going to playgrounds altogether.[32] It was a form of cultural resistance that mirrored the often successful efforts of adult workers to preserve their leisure time for "what they would."

The playground movement in Cleveland, Ohio, began modestly enough, with the Progressive mayor, Tom Johnson, calling in 1901 for the removal of all "Keep Off the Grass" signs from the city's parks. The philanthropic Cleveland Foundation, which was set up in 1914, quickly commissioned a survey of the city's schools. Next it turned to playgrounds, and in 1920, it issued its report on *A Community Recreation Program*. It was estimated that for every 1,000 hours a child spent in school, she or he spent another 1,800 hours in play away from school. In the aggregate, it was a substantial amount of time, and the report's introduction claimed that "just as the West was for long the great treasury of resources for the country, so spare time is the great treasury of unused or partly used resources of modern life." The director of the study was Leonard P. Ayers, and the director of the survey itself was Rowland Haynes, brought in from the Russell Sage Foundation. The first of a projected seven volumes was to be divided into reports on "Delinquency and Spare Time," "School Work and Spare Time," and "Wholesome Citizens and Spare Time." The volume most directly concerned with playgrounds was titled "Public Provision for Recreation" and was based on the assumption that, according to the report, "the most serious business in life for at least one-fifth of Cleveland's population is play." The authors made a great show of using scientifically generated statistics to demonstrate that, for example, children 4 to 10 years of age needed 87 square feet per capita for directed play and 174 for undirected play. By assuming that half the city's children, at any one time, were playing happily at home under parental supervision, the survey concluded that the city had sufficient play area and that therefore, any problems were caused by inadequate playground supervision, asserting that "play leaders must be generals of children." In Cleveland these leaders had originally been put through a course of instruction, but this step had later been dropped, presumably to save money. Female supervisors were paid less than men, but men were much more likely to quit their jobs. Twenty-seven percent of the leaders were judged to be only fair, and another 20 percent were ranked mediocre to totally unsatisfactory. In the bottom two categories, men outnumbered women two to one.

The physical condition of playgrounds was equally unsatisfactory. Only one of the city's seventeen playgrounds was ranked as acceptable, with the list of needs including resurfacing, "holes filled and leveled, proper fencing, adequate shading, more play apparatus, repair of equipment and more recreational supplies." As the report on delinquency had found, for large numbers of children the available playground technology, scientifically designed play programs and schedules, and the supervising play engineers could not match the appeal of the city's streets, lakefront, and industrial infrastructure. Ironically, to some extent it was adult technology that trumped mechanized play regimes. One boy reportedly stole some money, with which he promptly bought a "junior" auto, an Erector set, and a pair of field glasses to be used, he said, to "sight airplanes." A city bond issue was passed to buy equipment and upgrade apparatus, but the cost of playground maintenance was to come from the city's operating budget, and by the 1920s, they were once again run down and little used.[33]

Also by the 1920s, the extreme, and perhaps naïve, enthusiasm for Taylorizing playgrounds had begun to wane. This was no doubt due at least in part to the fact that the faddish aspects of Scientific Management, the idea that "the fundamental principles of scientific management are applicable to all kinds of human activities, from our simplest individual acts to the work of our great corporations," proved difficult to impose on the messy reality of social intercourse.[34] In part as well, American participation in the First World War from April 1917 occupied the time, effort, and ingenuity of play professionals. The construction of large military bases for the training of a newly recruited citizen army created centers of vice from which troops were to be diverted by organized sports and ample wholesome, recreation facilities. Lee Hanmer, for example, was made a member of the War and Navy Department committees on training camp activities, and in 1918, Gulick came out of retirement to chair the YMCA's International Committee on Physical Recreation of the War Work Council.

It was also true, however, that the aggressively Taylorist style of playground management had never gone unchallenged. A different philosophy, centered in Chicago rather than New York City, advocated a more creative, even aesthetic, approach to child recreation. Howard Bradstreet, the New York–based president of the Association of Neighborhood Workers, wrote that if in a playground "simple apparatus is erected, and a man placed in charge who understands boy nature and enjoys associating with them, the situation is changed. It is the personality which has the power of attracting and directing boys and which makes for the success of the playground, while the apparatus is but a device for keeping them occupied."[35] The case for a different approach was made by an unsigned editorial in the *Washington Post* in 1906, at the very founding of the Playground Association. It charged that

> while the leaders of the movement are busily advocating playgrounds for the children, they are at the same time determined to dictate to the little ones what they shall play and how they shall play it. . . . What healthy child, we should like to know, would give three straws for a so-called playground if there had to be some one there with authority to forbid this or limit that? Who wants to be watched, and lectured, and restrained and pulled and hauled about, and slapped, and straightened, and washed behind the ears, and tagged, and registered and kept account of and generally browbeaten from the time he or she enters the alleged playgrounds until an escape shall have been achieved? That isn't sport, or freedom, or enjoyment for youngsters. That's getting nagged, not to say spanked and put to bed.

It may have been a realistic preview of and preparation for working in the industrial world, but the strong advice of the writer was "Let the children alone!"[36]

In the decade following the war, the foundation of accomplishment laid down by the Progressive playground pioneers proved a durable platform for continued growth despite competing ideologies and the fluctuation of municipal budgets. In 1916, Carmelita Chase, who had trained as a teacher and worked with Jane Addams at Chicago's Hull House, married the patent lawyer Sebastian Hinton. At Hull House she had taken a two-year course in playground management and after her marriage she established a kindergarten in her home. Like the movement for playgrounds, that for kindergartens was a key part of the Progressive intervention in the experience of childhood.[37] In 1920, Carmelita Hinton's husband patented a metal climbing device marketed as a "Jungle gym," though later often popularly referred to as "monkey bars." It was claimed that Hinton's father, a mathematician, had built a similar structure out of bamboo to help his children grasp the concept of Cartesian coordinates. Hinton himself seems to have had a more realistic appreciation of its use. "I have found," he wrote in his patent application, "that children seem to like to climb through the structure to some particular point and there swing head downward by the knees, calling back and forth to each other, a trick which can be explained of course only by the monkey instinct."[38]

The continued interest in playground equipment was more than matched by the multiplication of sites for their use. Central Park in New York City had had no playgrounds until the philanthropist August Heckscher donated money for the first, which was opened in 1926 near the southwest end of the park.[39] Elsewhere in the country, municipally supported recreation thrived. The Playgrounds Recreation Association of America claimed, in 1927, that a million children were using such facilities every year and that 790 cities reported having more than 10,000 separate play areas "under leadership." It was asserted that in earlier playgrounds, children's "pursuits were entirely physical. But since the war playgrounds have been going in for other things as well. Music in particular plays an important role, and the drama follows close behind."[40]

Speaking in 1933 at a celebration of the opening of a playground in Central Park, Heckscher called for the building of playgrounds at every entrance to the park.[41] Under the "reform" administration of the city parks commissioner, Robert Moses, and with a significant infusion of money from the New Deal's Works Project Administration (WPA), eleven new playgrounds were opened in the city in 1936, including eight in Central Park. Another ten were planned along the park's margins. "They are equipped," it was reported, "with slides, swings, jungle gyms, play houses and sand boxes for small children."[42] In November of that year, the *New York Times* reported that two more playgrounds had been opened, bringing the total to 188 of new playgrounds built since Fiorello La Guardia had become mayor in 1934. One of the new facilities was the first of twenty-four only recently authorized. This site had "swings, seesaws, sand tables, playhouses and a jungle gym for younger children, as well as handball courts, play apparatus and a large open play area with a softball diamond for the older children. There will also be a comfort station for men, women and children."[43]

Beginning in the 1930s, the subsequent history of playgrounds in the United States saw an effort by a number of landscape architects to introduce the modernist vocabulary of new materials and free-flowing spaces in place of the traditional flat, paved grounds with steel play equipment. This was, in a sense, a harkening back to the more environmentally oriented Chicago and Boston approaches to playgrounds. In 1933, the Japanese American artist Isamu Noguchi proposed to build what he called a "Play Mountain" in New York City, which would use his technique of "molding earth" by tilting and excavating a city block. The sculptor

This girl swinging safely in a playground represents
the ideal not only of child play reformers but also
of the manufacturing firms that were prepared
to supply the desired playground equipment.
Funful. The Playground Equipment Line. Charles J.
Jager Co., 1920 Trade Catalog, Smithsonian
Institution Libraries, Trade Literature Collections.
Courtesy of the Smithsonian Institution Libraries,
Washington, DC.

himself claimed that Play Mountain "was the kernel out of which have grown all my ideas relating sculpture to the earth. It is also the progenitor of playgrounds as sculptural landscapes." The scheme, however, was turned down by the same New York parks commissioner, Robert Moses, who was so enthusiastically building traditional playgrounds.[44]

Noguchi went on to reinvent swings and a spiral slide, but was rejected for a project in Hawaii and again in New York, where it was considered too dangerous. This last unrealized project, called the Contoured Playground, had no equipment at all but was characterized by an undulating landscape. "I felt obliged," Noguchi wrote, "to answer the dire warnings of the danger to which I would expose small children with my play equipment and so designed a Contoured Playground. This would be proof against any serious accidents, being made of entirely earth modulations. Exercise was to be derived automatically in running up and down the curved surfaces. There were to be various areas of interest, for hiding, for sliding, for games. Water would flow in the summer."[45] It was turned down in 1941, in part because of America's increasing involvement in World War II.

After the war the new ideas came from Europe, particularly Scandinavia and the Netherlands, where the wholesale devastation of cities made room for a rethinking of playgrounds. The first Adventure playground was actually organized in Copenhagen during the German occupation. A professional supervisor, present only to give advice, watched children build with rubble and a few hand tools. As one student of the subject has written, "these playgrounds seduced kids by making them believe that they had entered a forbidden zone, gaining access to materials that appeared inappropriate for play, and tools that usually were reserved for adults."[46] Through the work of Aldo van Eyck, Amsterdam became a center for such playgrounds, and, under the leadership of Lady Allen of Hurtwood, Adventure Playgrounds came to Great Britain in the 1950s.[47]

In the United States after the war, those thinking about playgrounds fell into one of two groups, coming from either the recreation movement or the art world. The former thought in terms of individual pieces of equipment (still the traditional sandboxes, slides, seesaws, and swings) and programmed activities (such as folk dancing and dramatics) under the direction of supervisors, of whom there were some 50,000 in 1948. It was a split reminiscent of the 1906 New York–based champions of scientific play and the Chicago-based advocates more free-style activities. Significantly, before the war the Playground Association had changed its name to the Playground and Recreation Association of America, underscoring its de-emphasis on Taylorite play and reorientation toward organized activities.

Alternatively, the art world emphasized the connection between playgrounds and sculpture. Once again a plan by Noguchi was thwarted by Moses (who allegedly called it a "hillside rabbit warren"), this time for a playground for the new United Nations site. A model of the proposed construction, described as resembling "in part a contour or relief map, indicates that much of the equipment would be built like sculpture. There are tunnels, climbing areas, some of curved shapes, others a sort of step-like mountain of brightly colored triangles, spiraling paths and a sloping area of multiple slides in addition to jungle gyms, a framework for swings, sheltering wall, a wading pool, etc."[48] Noguchi revealed his intentions by noting that "a jungle gym is transformed into an enormous basket that encourages the most complex ascents and all but obviates falls. In other words, the playground, instead of telling the child what to do (swing here, climb there) becomes a place for endless exploration, of endless opportunity for changing play."[49]

An intervention by New York's Museum of Modern Art proved futile, and the argument that this project had "possibilities of stimulating the child's sense of space and form through a playground designed as architectural sculpture" fell on deaf ears: the playground design that finally was approved was from Moses' own department. MOMA, however, went on to exhibit the model of Noguchi's playground in its Young People's Gallery in 1952. Philip C. Johnson, the director of the museum's Department of Architecture who eventually became one of the country's most famous and influential architects, claimed that Noguchi's model was meant to demonstrate ideas rather than prescribe a solution. "The playgrounds authorized by our parks and recreation authorities," he admitted, "are characterized by skillful engineering and landscaping, careful planning and well-conceived commercial equipment. Although such playgrounds meet the requirements of safety and supervision while withstanding hard use," he added, "there have been few efforts to go beyond solving mechanical and functional problems."[50]

Two years later, in 1954, MOMA was again the venue of a playground exhibit, inspired by Noguchi's UN model, and this time co-sponsoring a "Play Sculpture Competition." The other two sponsors were *Parents Magazine* and the commercial firm, Creative Playthings. Winners of the design contest, which had been announced the previous year, were promised cash prizes, Creative Playthings would manufacture the winning designs, and the museum would be given ownership of the designs themselves.[51] A critic writing in the *New York Times* called it a "strange and wonderful world of color and shapes," noting that "in general form and concept the 360-odd entries follow the recent directions in the playground field. There are geometric, constructive shapes, from cubed jungle-gyms to elaborate polyhedrons; there are plastic or sculptural shapes; there are animals, realistic and fantastic; there are variations of conventional play apparatus such as swings and slides; and there are variations on the 'junk-yard' theme of pipe sections and other oddments which have special attractions for the young." The critic was less enthusiastic about the entries' artistic elements, lamenting that "only occasionally does an entry have the authority of form and the integrated concept of Noguchi's project for a UN playground," and that they lacked "cognizance of mass production possibilities—which a Charles Eames would have brought to the problem." He concluded by observing that to visit the show "might even be a civic duty. Perhaps then the next time New York has an opportunity to have a playground as fine as the one Noguchi designed for the U.N. site, the citizenry will prevent a Moses 'thumbs-down.'"[52]

An article in *Art and Architecture* viewed the competition as a means to help overcome a most unsatisfactory status quo. "Playgrounds for children," it affirmed, "are an essential part of modern city planning, and the quality of their play equipment is of vital importance. However, the cement-floored, wire-fenced patches of recreational areas set aside in city parks and schoolyards, and fitted with monotonously identical metal constructions for physical exercise, are cogent proof of how inadequately we have estimated their importance in our communal life."[53]

The critical role of Creative Playthings signaled the growing importance of commercial manufacturing firms in the playground-supply business. The firm had its roots in a shop established in 1945 that carried playthings from a variety of manufacturers. The company soon added some of their own design, most famously a line of hardwood blocks. The owners, Frank Caplan and Bernard Barenholtz, sought the help of the Museum of Modern Art to bring a modernist sensibility into the aesthetics of play, and in 1949 succeeded in commissioning the architect Marcel Bruer to design hollow cubes for the playroom. In 1951, they set up their new firm Creative Playthings to commercialize their new designs. Caplan hoped to create a "complete playground that could be planned by a 'small panel of key designers' and then be marketed to public institutions such as housing authorities, park departments, and school boards."[54] He managed to recruit the sculptor Robert Winston, held talks with Noguchi, and on a trip to Europe he contacted Egon Moller-Nielson, a Dane who had actually built pioneer abstract playground pieces in Sweden. Even Henry Moore signed on, though apparently he never actually produced any designs for Caplan.

At the annual meeting of the National Recreation Congress, held in Philadelphia in 1953, Creative Playthings "attracted keen interest," according to the *New York Times*. Their exhibit of the company's products included "concrete mountains, with built-in caves, slides and climbing arrangements, a metal climbing gymnasium, a fiberglass turtle with hollow sculptured shell to use for a tent and a see saw in the shape of an airplane." The concrete and stone slide was a modernist creation clearly showing the influence of the sculptors working with the firm.[55]

During the 1950s and '60s, the company was quite successful, becoming a major source of toys for some 60,000 of the nation's nursery schools and kindergartens. Equally important, it became the vector by which middle-class parents, already wearing Marimekko fabrics, sitting on Aalto stools, and dining from Arabia dinnerware, could introduce modernism into their children's nursery. Its offerings numbered over a thousand, ranging from "a two ounce teething ring to a four-ton piece of playground equipment."[56] An early success of Creative Playthings' Play Sculptures division was being chosen to supply equipment for the American National Exhibition held in Moscow in July 1959. The playground, rather than advertising itself as a community resource, stood next to the model middle-class home outfitted with all the latest, modern appliances and conveniences—where President Richard Nixon and Soviet General Secretary Nikita Khruschev had their infamous kitchen debate.[57]

In fact, the playground was perhaps even less "typical" than the domestic dream next door. Both were aspirational, but the Play Sculptures represented a high end of American playgrounds at mid-century, not their common design. Nevertheless, in the late 1960s, there was a brief efflorescence of landscape-inspired design in New York City, led by the landscape architects Richard Dattner and Paul Friedberg. The latter's Jacob Riis playground on the Lower East Side was described as "a profusion of nooks, crannies, angles, pyramids and mounds of richly textured rock. There is a maze, another splashing pool with a miniature waterfall, a tree house. It is a vast sandbox with all the pieces set inside . . . [designed to produce] 'a continuous play experience'"[58] Dattner created an Adventure Playground in Central Park, and in 1967 the Museum of Modern Art once again became involved, cooperating with the New York City Housing Authority and the Park Association of New York City in designing a "complex of vaults, with spaces to climb in, slide over and hide in."[59]

After being prevented from constructing several playgrounds in New York City, the Japanese-American sculptor Isamu Noguchi was finally allowed to design one in Piedmont Park in Atlanta, Georgia, in 1976. The modernist shapes and bold colors of the slide and swings are his signature. With the permission of the photographer, Clare Mullins.

This is also a good description of the modular "PlayCubes" patented by Dattner in 1969. Made of fiberglass, they were hollow with openings on all sides and meant to be fitted together to create an endless variation of shapes and passageways. He claimed that they were particularly useful for the new "pocket parks" that were being "carved out of empty lots, oftentimes of irregular shape, . . . [and with] little uniformity between different play areas." He envisioned that they would also be useful for "homeowners in yards."[60] In 1974 he designed a variation called "Habitot," to be used by very small children.[61]

Noguchi himself made one more attempt to build a playground in New York City when he was commissioned, along with the architect Louis Kahn, to build the Adele Levy Memorial Playground along Riverside Drive. Lasting from 1961 to 1966, the design went through six iterations and the site was shifted three times. Although originally having the support of the city administration, it was opposed by a group of local residents who eventually succeeded in blocking it completely.[62] Charles Starke, the city's recreation director, was less than enthusiastic as early as 1962, in part at least because in his opinion, "we haven't seen any new type of equipment that in our opinion appeals to children more than what we've been using."[63] Noguchi was finally able to realize one of his "playscapes" in Atlanta, Georgia, in 1976. Building on a site once used for the 1895 Cotton States and International Exposition (which had included a Phoenix Wheel, scenic railway, and Shoot the Chutes), Piedmont Park allowed Noguchi to create a playground designed to familiarize children "with shapes, colors and textures."[64]

In 2007, New York announced that the architect Frank Gehry would, as a newspaper put it, "bring his daring deconstructionist aesthetic to the monkey bar and seesaw" with a new playground at Battery Park.[65] In 2010, an "Imagination Playground," the design of architect David Rockwell, opened at the South Street Seaport. With almost no fixed equipment (certainly not the so-called four 'S's of older playgrounds—swings, slides, sandboxes, and seesaws), the site featured Rockwell's "loose parts," large "lightweight blocks made from bright-blue molded foam" which came in various shapes. Wooden wheelbarrows, tires, plastic barrels, and similar parts were also available. Such high-end facilities were expensive, of course, made more so, in the case of the Imagination Playground," by the presence of "Play Workers."[66]

Even in their heyday these "playscapes" were far from typical. The American media giant CBS's (Columbia Broadcasting System) purchase of Creative Playthings in 1966 was followed by a wave of consolidation in the industry. Some older companies survived: Miracle Recreation Equipment Company had been founded in 1927 to manufacture merry-go-rounds, and at the end of the century was offering eight different modular "play systems" as well as more traditional slides and swings.[67] However, by the late '70s four large companies dominated, selling so-called post and platform combinations made of heavy logs. Since wood structures and fiber nets eventually deteriorated, however, brightly colored polyethylene became more prominent. The apotheosis came with the more than 8,000 indoor and outdoor playgrounds at McDonald's fast food restaurants. Basically post and platform creations in garish colors, designed to look larger and bolder than they really were, they offered children the play equivalent of burgers and fries: kids could climb up, cross over, and go down—then start over. Not surprising in the light of its ubiquity, the McDonald's playground threatened to become the American standard for play facilities.[68]

By the start of the twenty-first century, however, the popularity of "Adventure" playgrounds was long over. In 1999 it became evident that "these adventure playgrounds represent a period in time—much like the monkey bars of the 30's. . . .

Play equipment rooted in 60's idealism . . . [was being] replaced by the flashier mass-produced equipment of the 90's." Michael Gotkin, a landscape designer, lamented that "there was a brief window when artists, architects, sculptors and parents got together and created a playground revolution. It was so recent but is so entirely gone. You don't really design playgrounds now. You order them out of a catalogue." The problem, according to Friedberg, was the old issue of risk: "we are protecting the child too much," he complained.[69]

The recurring issue of safety, which was instrumental in getting the playground movement started at the turn of the twentieth century, was never far from the surface of public concern. Surprisingly, even some of the judges of the MOMA/Creative Playthings competition of 1954 were said to have privately referred to some of the entries as "101 Ways in Which to Kill Your Child."[70] Although more than a dozen Adventure Playgrounds in the European style were established, their apparently chaotic design led the American insurance industry to ignore their excellent safety record and make them virtually impossible to continue operating in the United States. In conjunction with the legal profession, the two conjured up visions of parental litigation that few public agencies could afford to ignore. When Lady Allen toured the country in 1965, she called the country's playgrounds "an administrator's heaven and a child's hell." The three principles in their design seemed to be cheap cost, low maintenance, and avoidance of litigation. Americans, Lady Allen remarked, "seem to be terrified of risks—they are dogged by fear of insurance claims resulting from accidents in public playgrounds. I've never seen anything like it." Her own position was quite different: "I think it's better to risk a broken leg," she said, "than a broken spirit."[71] The widespread concern, however, was not entirely misplaced. In the 1980s, after paying nearly $10 million in claims, the Chicago Park District removed its spiral slides, high monkey bars, and merry-go-rounds from its sites. New York City got rid of seesaws.[72]

The concern for playground safety was nestled in a much larger national obsession. Partly in response to well-documented instances of corporate disregard for consumer welfare, a Consumer Product Safety Act was passed; in 1973 the resulting Consumer Product Safety Commission began to target areas of particular danger using data collected from 119 hospital emergency rooms. Playground equipment quickly emerged as a prime offender, and in 1981, the Commission issued guidelines for public playgrounds. One student of the subject has argued that, despite the wave of privatization and deregulation that subsequently swept over the country, playgrounds remained targets of sometimes draconian concern.

Since 1991, the National Recreation and Park Association has backed a National Playground Safety Institute which runs two-day courses in the subject. Graduates become Certified Playground Safety Inspectors who conduct safety audits and are empowered to close down any project that involves play.[73]

Despite all this, in 2001 *Scientific American* reported on a study in the journal *Ambulatory Pediatrics* which claimed that, while a quarter of all emergency room admittances of children and adolescences were because of accidents, of these only 5.3 percent occurred in playgrounds, and only 9 percent of those occurred in *public* parks or playgrounds. Nevertheless, the story was headlined "More Children Are Seriously Injured on Playgrounds than in Traffic." The following year the Ralph Nader–inspired Public Interest Research Group (PIRG) claimed that "on average, 17 children die each year playing on playgrounds. Many of these deaths and injuries can be prevented if playgrounds, from equipment design to surfacing content to the playground's layout, were designed with safety in mind."[74]

The perhaps inevitable result of this kind of assault was the continued stripping of traditional play apparatus from sites across the country. A count in 2004 showed that, while 94 percent of playgrounds had slides, only 13 percent had seesaws, 7 percent had merry-go-rounds. A 2008 report found that, "though smaller swing sets still dot backyard lots, the glorious 16-foot models that generations of kids grew up loving and fearing, are vanishing. Disappearing as well: traditional seesaws, sandboxes and those merry-go-rounds from which you tried to throw your little brother. (Somehow he always hung on.)"[75] The result led one critic to declare that childhood was "becoming undermined by risk aversion."[76]

That the new culture was not entirely hegemonic was suggested by the results of a request, early in 2009, by the *San Francisco Chronicle*'s blog for parents, the "Poop," that its readers identify "public places that are good for kids—a.k.a., playgrounds." A list of six was published: one, Howarth Park in Santa Rosa, was praised for having "huge, complex climbing structures, lots of water features for hot summer days, volcanoes, Native American dwellings and a mini 'Tech Deck' skate park." Ryder Park in San Mateo was described as a "big kid play area [with] . . . a 12-foot-high rope climbing structure, [and] funky art that doubles as play equipment." And one reader identified the Koret Children's Quarter in San Francisco's Golden Gate Park. Singled out for praise was the fact that "the architects weren't complete wussies when designing the playground: They created several semi-dangerous additions, including a concrete slide and a giant rope-climbing structure."[77]

The story of American playgrounds over the past century reveals a dense tangle of technology, politics, economics, class bias, professional aspiration, and cultural idealism. Technology is both cultural process and cultural product. During the Progressive Era, from roughly 1890 to 1920, the ideologies of science and reform drove the subject of children's play as they did so much else in the United States.[78] By mid-century, however, as the cultural historian Nicholas Sammond has argued, science and reform had been replaced by the new media and child consumerism.[79] He was writing of television, of course, but the advent of the Internet and its attendant technologies only strengthens his case. Where children in playgrounds *produced* play, at Disneyland they *consumed* play. While playgrounds still existed in the thousands, they had lost the Progressive cultural imperative that had given them such a prominent place in the cultural construction of childhood—and such a clear claim on the public purse.

The classic schoolyard playground, the regimented Taylorite sites, the European Adventure Playground, the Noguchi playscapes, the McDonald's play areas, and the sanitized and neutered modern playgrounds each represent a menu of philosophy, design, and equipment from which social intention can, and did, in Huizinga's phrase, create "temporary worlds within the ordinary world, dedicated to the performance of an act apart."[80]

Chapter Three

From Pleasure Gardens to Fun Factories

Nothing to depress or demoralize.

—EUCLID BEACH PARK, CLEVELAND (C. 1901)

Children's Fairyland, in Oakland, California, was built in 1950. Fondly described as "low tech," and in 2010 as a "throwback amusement park," it is accessible only to children and any adults they might care to bring along with them. Willie the Whale and Oswald, a bubble-blowing elf, greet visitors at the gate, and a small train winds its way through grounds dominated by child-size buildings and exhibits. A theater at one end was, in 2010, staging "The Little Engine That Could." Walt Disney reportedly visited Fairyland before he built his own Disneyland, and he is said to have hired away the executive director and puppeteer as well.

One technological innovation of the park was the creation of a key that children could use to activate talking "story boxes" located throughout the grounds. The idea was the creation of Bruce S. Sedley, a children's program host on local television. The keys were a great success (the earliest keys still unlock the story boxes), and Sedley went on to patent a magnetic card-swipe device such as that now widely used for hotel doors.[1] Part of the charm of the low-tech Fairyland, however, lies in the very fact that it stands in stark contrast to most amusement parks in the United States.

Commercial amusement parks in the United States, beginning at least with New York City's Coney Island in the 1870s, were entertainment venues crowded with technological wonders, from roller coasters to monorail trains, "burning" buildings to the Ferris Wheel, invented in 1893 by the engineer George Ferris for Chicago's Columbian Exposition and now a marker of many large European cities. But the rides and spectacles, which promised the illusion of danger (what historian Arwen Mohan has called "The Commodification of Risk") and often mimicked the technologies of society at large, were only a part of the experience. Besides having rides that produced thrills through the illusion of danger, amusement parks also, importantly, were sites of heterosocial interaction where (especially young) men and women could mix promiscuously, clinging to each other on wild rides, watching each other caught literally off balance in fun houses, and simply spending time together without direct supervision. Both the technology of the rides and the intimacy of the interactions were thrillingly modern.[2]

Amusement parks were wildly popular in the United States from the late nineteenth century on. When Luna Park was constructed near Cleveland, Ohio, in 1904, it was the thirty-fourth such project undertaken by Pittsburgh's Ingersol Construction Co. These parks were embedded in networks of transportation and power technologies, linked to city centers by streetcar lines, thus providing both a destination for riders and a daytime use for the electricity produced by new power stations that needed to balance their loads. The original Disneyland was carefully placed adjacent to a not yet finished Los Angeles freeway interchange. Indeed, much of the success of Disneyland can be traced to the confluence of the automobile, the freeway, television, and the "imagineering" of Disney's designers. In recent times, these sites have been designed and constructed as efficient producers of pleasure. The later Disney World is a fantasy delight for the consumers of pleasure, but it sits above another, more brutal and rational world of underground utilities, warehouses, and passageways. As Michael Sorkin has noted, Disneyland reversed the older world's fair ideal of a celebration of production to a production of celebration. The theme parks are indeed sites where "nature is appearance, machine is reality."[3]

American amusement parks could trace their origins back, in part, to the Pleasure Gardens of eighteenth-century England. Following the Restoration of the monarchy in 1660, with Charles II coming to the throne, pleasure began to be an acceptable public feeling once again. Gardens, privately owned but open to the public, began to appear, offering upper-class urban dwellers the pleasures of fine art and classical music (Mozart made his English debut in such a place), all set in enchanting and exquisite landscapes of fountains, promenades, vistas, and artfully designed ruins. During their heyday in the 1740s and '50s, they featured concerts, balls, masquerades, public meals and, perhaps most spectacularly, fireworks displays. By the Victorian era, however, most such pleasure gardens had lost what one scholar has called "their Arcadian chic" and had become sites of family entertainment.[4]

One of the best known and earliest of these sites was the Vauxhall Gardens in London, which lasted from 1661 to 1859. One reason for its great success, as proved true years later with Coney Island and other American amusement parks, was the opportunities it provided for the sexes to mix informally without many of the restraints of polite society. Another was that at Vauxhall, in the mid-eighteenth century at least, for the admission price of one shilling (a not inconsiderable sum) anyone could mix with their betters and even, with luck, catch sight of the Duke of Wales ("ground landlord" of the site). A drawback, at least for some, was the large number of prostitutes who took advantage of the informality and the presence of wealthy men to ply their trade.

The experience of a visit for most was enhanced by the dramatic, and even a bit dangerous, crossing of the Thames River to the landing at the Vauxhall Stairs on the Surry Bank. From there the visitor could move forward toward the orchestra building, where music was a major attraction of the park. After a concert and walk through the woods and gardens, a meal was served, at about nine in the evening; as the darkness fell, servants, who had been stationed around the property, on a set signal lit more than a thousand oil lamps, providing some of the drama that electric lights would create a century and a half later. By the time the park closed in 1859, a railroad ran past the entrance, which could also carry London residents to further amusement centers, often along the seacoast and built out over the water on piers such as at Brighton and Blackpoll.[5]

Denmark's famous Tivoli Gardens marked a later, and still popular, evolution of the pleasure garden. Georg Carstensen, born in 1812, was the son of a Danish diplomat and had a career in the military Royal Guards. In 1839, he moved to Copenhagen where he established two periodicals which he promoted by organizing illuminations: with drinks, music, and a fireworks finale. In 1841 Carstensen applied to King Christian VIII for permission to establish his "Tivoli and Vauxhall," which he then opened in 1843. His plan for Tivoli, as it came to be called, was based on the pleasure gardens he had seen in other parts of Europe. It was to include a bazaar for Danish and foreign products, a restaurant, a music pavilion, a theater for dance and drama, and so forth, of course featuring fireworks. Also to be included were various rides, the first two of which were a carousel and a so-called switchback, which was actually a long slope down which people rode. Instead of horses, the carousel carried small railroad cars (four years before Denmark's first rail line was constructed). The switchback, or Rutschbane, offered a seven-second ride to the bottom. Eventually Tivoli had "30-odd games," four "slot machine halls," and 24 rides, including the 1914 wooden roller coaster, the oldest still in use. Its cars are pulled to the top, and then descend by gravity, reaching a speed of 50 kph.[6]

Ironically, it was the technological expansion of the Industrial Revolution that helped bring Vauxhall to an end but also helped Tivoli survive. The democratization of society created a mass market for the kind of amusements that could provide relief and distraction from lives played out in factories and shops, offices and tenements. The great insight provided by Vauxhall and perpetuated today in a thousand shopping malls was that crowds of people themselves provide the spectacle that attracts and excites those same people. The new amusement parks, such as those that sprang up in the United States, were also filled not only with excited crowds but scores of machines, built of steel and powered first by steam and then electricity, which threw people together promiscuously, their excitement heightened by speed and intimacy as well as a carefully engineered illusion of danger. Another irony is that the new amusement parks created a faux replica of the very technological world the crowds were presumably seeking to escape for a brief while.

Changes in transportation technology were critical to the way in which amusement parks developed in America. Railroads, ferries, and subways increased the range of urban mobility and all were used to move crowds to such destinations as Coney Island. The first electric streetcar line was built for the Berlin Exhibition in 1879. Frank J. Sprague, who had worked briefly for Thomas Edison, built the first American line in Richmond, Virginia, in 1887. He lost money on the venture, but his later efforts were more successful and streetcars spread across the country. Within a few years, more than a billion dollars had been invested in these lines, and more than three hundred million in the lighting plants that supplied them with electricity.[7]

The Cosmopolitan magazine in 1902 carried an article titled simply "The Trolley-Park." "The fact is," it asserted, "the street and suburban railway companies, realizing the profit arising by catering to the pleasure of the masses, have entered into the amusement field on an extensive scale." According to the author, "trolley-parks" were to be found "in the outskirts of nearly every city in the land" and often as "centers of recreation for clusters of small communities which may be linked by the electric current."[8] Streetcar systems had been set up to transfer "the throngs of toilers to and from store, office and factory, and forming a means of transportation between the different parts of the city." The lines were often extended into the nearby countryside, where land developers built "streetcar suburbs" to cater to a growing middle class. Even city dwellers not living in the suburbs took advantage of the new lines to cool off during summer evenings: "People of all classes availed themselves of the opportunity to get a breath of fresh air and pass the long evenings enjoying the 'trolley-breeze,' for a rapidly whizzing trolley-car can stir a breeze in the stillest night of midsummer."[9]

Recreational trolley-riding was an unexpected way to increase revenue for trolley companies, but an additional and even more important advantage was the opportunity to better balance the electrical load available. All trolley companies needed electrical generation facilities to provide current for their cars, and therefore were the subject of power as well and transportation concerns. It was also true that power companies had to try to use as much electricity as they could produce, otherwise expensive capacity would stand idle. Use varied throughout the day and night, and from season to season. The more, and more varied, the uses and the more evenly demand was spread around the clock, the more profitable the company became. Trolley parks helped in both respects.

The numbers of revelers were impressive: "On a holiday one may see more than fifty thousand people gathered in some of the more extensive trolley parks, . . . listening to band-concerts, watching or taking part in ball games, boating on the lake and river," or enjoying the numerous "amusements" available. Some were comparatively simple, like boys paying "their pennies for the privilege of hurling a ball at the shining face of a negro, who usually manages to duck just in time to avoid disaster, but not too easily to destroy the hope of nailing him next time. And it is wonderful with what savage zest the small boy will take aim until his last penny has been handed across the board." Other amusements were more elaborate, since "the inventive genius has made a study of diversions for the trolley-park, which perhaps represents a greater variety of popular amusements than any other resort."[10]

As amusement parks were constructed across the United Sates, they were filled with "rides." Probably the most iconic, and often the first ride built in a fledgling park, was the carousel or merry-go-round. The classic carousel, with its wonderful carved horses and other animals, had its roots even deeper in European culture than the amusement parks themselves. The word "carousel" had long meant a tournament of knights, but also referred to a game played by Turkish and Arabian horsemen to sharpen their fighting skills. In some cases it referred specifically to a training exercise which involved putting a spear through rings attached to revolving arms—perhaps the precursor of the rings the modern

Well-dressed merry-makers used the Los Coches flume trestle in
Southern California, built as part of a water supply scheme, as a
pleasure ride far from the nearest amusement park. The repurposing
of technologies is a common theme in their social usage. *Scientific
American*, 62 (March 15, 1890), 161.

carousel riders attempted to grab as they passed by. In 1673, a writer referred to "a new and rare invencon knowne by the name of the royalle carousell or tournament being framed and contrived with such engines as will not only afford great pleasure to us and our nobility in the sight thereof, but sufficient instruction to all such ingenious young gentlemen as desire to learne the art of perfect horsemanshipp."[11]

Beginning at least in the eighteenth century, carousels were manufactured and operated at fairs around Europe and England. Such early devices had no platforms, but consisted of horses suspended from a central pole. As the pole turned, centrifugal force pulled the horses and riders out into a widening circle. Called "flying horse" carousels, these were the first kind used in the United States, with one recorded in New England in 1800.[12]

In 1837, Michael Dentzel was traveling around southern Germany, bringing his portable hand-carved carousel to fairs and markets. His sons followed in the trade, and in 1864, his second eldest, Gustav, like his brothers, emigrated to America, bringing with him by sailing ship what is said to be the first proper amusement park carousel in the country. The company they established, the Dentzel Carousel Company, hired other German and Italian wood carvers to craft their handsome animals. Indeed, in addition to horses, the Dentzels adorned their carousels with cats, rabbits, giraffes, lions, tigers, deer, goats, pigs, donkeys, kangaroos, buffaloes, ostriches, dogs, and more. Their machines between 1837 and 1928 were sometimes as large as 54 feet in diameter, with 72 animals and four chariots.[13]

Gustav's two sons continued the business after his death in 1908. William died in 1928, by which time his brother Edward was in southern California setting up and operating carousels. The latter eventually left the business and began building homes in Beverly Hills, where he was a city councilman and then mayor from 1933 to 1947. Edward's son William H. Denzel II was a lawyer who inherited the family passion for carousels, building a line of children's machines, complete with their trademark mirrors, artwork, lights, and band organ music. William II died in 1991, but William H. Dentzel III and his three children continued the business. Following the remarkable family tradition of craftsmanship and innovation, they began to produce solar-powered carousels, like that for the Solar Living Institute in Hopland, California, in 2005. At the other end of the technical spectrum, they produced a "rope-pull human powered" flying horse carousel as part of a community project in Chiapas, Mexico. They also make foot-pedal powered devices.[14]

Charles W. Parker, who was, in 1918, rather grandly called "the world's Napoleon in the manufacture of amusement devices," appears to have been more of an exemplary American entrepreneur of his time. Born in Illinois in 1864, he moved with his family to Kansas in 1869. He apparently found his métier in 1882 when he borrowed money to buy a shooting gallery. The next year he built a striking machine (or "High Striker") which encourages young men to pound a small platform to send a ball up a narrow wall to ring a bell. Parker traveled from town to town with these two amusements and in 1892 added a used carousel. By 1898, he had modified its design to the point where he could claim to have an essentially new ride, which he called a "Jumping Horse Carry-Us-All." In 1902, he formed the C.W. Parker Amusement Co. His traveling company, called Parker's Greatest Show after 1916, required a train of thirty-five railway cars to carry it from town to town. His manufacturing company meanwhile was called "the largest in the world devoted exclusively to the manufacture of amusement devices." It turned out a range of rides, but concentrated on the "Carry-Us-Alls and the thousands of wooden horses, mechanical organs, and other devices required in the show business."[15]

Besides the Dentzels, other European craftsmen emigrated as well, and between 1880 and 1930, the United States had sixteen carousel and carving shops, six in Brooklyn, New York alone. The technique of building the spectacular horses and other animals was similar in all the shops: pieces of wood were glued together to build up a thick, box-shaped block from which the animal's body was then carved. The head and neck were carved from similar but hollow boxes. The sections were carved separately, then fastened together with glue and dowels. In 1914, the Moore Carving Machine Co. introduced its Lochman carving machine, which operated somewhat like Blanchard's copying lathe, using a master copy of each part to produce four others at the same time. Eventually aluminum animals were produced, and late in the twentieth century some were made of fiberglass. The Dentzels continued to carve wood.[16]

The earliest carousels were powered by animals or people, and in some cases children, who were allowed a free ride for their efforts. The use of steam power seems to have come somewhat late, with the first known instance occurring in England in 1861, and eventually many were driven by electricity.[17] With this added power, carousels could be made larger and heavier, with more elaborate machinery and up to one hundred animals in five concentric rows. Efforts to introduce more realism, or at least motion, led to using an overhead crank system to make horses "gallop." At the beginning of the twenty-first century, the National Park Service estimated that there were 135 still functioning antique carousels in the United States.[18]

Like carousels, what later came to be called ferris wheels date back several centuries. In the 1620s, the English traveler Peter Mundy came across something like a fair being held in a small town in Turkey. Among several amusement rides, he wrote, was one in which two tall poles, joined at the top by an axle, supported two large wheels between which hung seats. As he described it, "Children sitt on little seats hung round in several parts thereof, and though it turne right upp and downe, and that the children are somtymes on the upper part of the wheele, and sometimes on the lower, yet they always sit upright."[19] By 1728 a similar device, called the "ups-and-downs," was sketched at St. Bartholomew Fair in England, and another sketch, from 1860, depicts a sixteen-passenger "pleasure wheel" in France. Both of these last were turned by hand cranks. It is claimed that Antonio Maguino erected a wooden hand-turned wheel at his picnic grounds at Walton Spring, Georgia, in 1848. Wooden wheels were fabricated for sale as early as 1870 by Charles W. P. Dare of Brooklyn (who also made carousels), and, in the 1880s, the Conderman brothers of Indiana were selling wheels made from metal pipes and carrying ten double seats.[20] In 1891, William Somers, of Atlantic City, New Jersey, applied for a patent for his "new and useful Improvements in Roundabouts." The patent was granted in 1893, by which time he had built three fifty-foot wheels; one in Asbury Park, one in Atlantic City, and the third at Coney Island. It was the Atlantic City wheel that apparently served as a model for what would become the famous Ferris Wheel.[21]

The planners of Chicago's Columbian Exposition, designed to celebrate the 400th anniversary of the "discovery" of America, wanted a structure that would capture the imagination of fair-goers as had the observation tower of the Philadelphia Centennial Exposition and, most notably, the great tower built by the French engineer Gustav Eiffel for the Paris Exposition of 1889. George Ferris, an engineering graduate of the Rensselaer Institute of Technology and a Pittsburgh bridge builder, decided on a giant observation wheel, 250 feet in diameter and supported by two towers, 140 feet high. Attached to the wheel were to be 36 cars, each designed to carry 40 passengers—a total of 1,140 people for each ride. In March 1893, five trains, each with thirty cars, arrived in Chicago with the steel parts for the wheel. It was still unfinished when the fair opened in May. After erecting the structure, attaching the wooden cars, and finally testing the two 1,000-horsepower steam engines (one to drive it, the other for emergencies), the

The original giant Ferris Wheel at the World's Columbian Exposition in Chicago, 1893. Its monumental proportions were chosen in a deliberate attempt to mimic, and outdo, the tower built by that other engineer, Gustave Eiffel, for the 1889 world's fair in Paris. *Scientific American*, 69 (July 1, 1893), 1.

wheel was opened on June 21. At 50 cents a ride, ten times what a carousel would cost, it was a huge success. Three thousand electric light bulbs kept it operating into the night, and by the end of the fair, 1,453,611 paying customers had ridden the wheel. After the fair, the ferris wheel, as it was now called, was taken down and moved to another site in Chicago. Finally, it was moved to the grounds of the Louisiana Purchase Exposition in St. Louis in 1904, but after the expo closed, the wheel was destroyed.[22]

A month after the Columbian Exposition ended, the president of the Civil Engineer's Club of Cleveland declared, "Gentlemen, the Ferris Wheel has excited much interest all over the world and among all intelligent people, and especially among engineers of all classes."[23] One was a retired British naval officer, W. B. Basset, who decided to make a second career of building giant wheels; in 1895, for the Earl's Court, London, Oriental Exhibition, he erected one which, at 270 feet diameter, was 20 feet larger than the one Ferris had created. Basset built another, of 200-foot diameter, for the seaside English resort of Blackpool, in 1896, one of 197-foot diameter for Vienna in 1897 (still standing in 2014), and finally, a 240-foot diameter wheel for the Paris Exposition of 1900.[24]

While Basset built large, William E. Sullivan thought smaller; specifically, he sought what later came to be called transportable ferris wheels that could be used by carnivals, for example, which traveled from town to town. Like so many others, Sullivan had ridden the ferris wheel at the exposition and been inspired. He owned the Eli Bridge Company in Roodhouse, Illinois, and, with the machinist James H. Clements, constructed their first wheel in 1900. In 1906 he incorporated his company (still operating in 2011) and began to manufacture his wheels, "The one SAFE, SANE Wheel, quick to take down, quick and easy to erect. Not a bolt to remove."[25] In 1905, he received a patent for a "Swinging Seat for Use in Amusement Wheels," which was intended to "provide a seat which can readily be folded for protection and convenience in moving and which provides a safety-bar to fold in front of the occupant when the wheel is in operation."[26]

Over the years, variations on the basic design of the ferris wheel continued to be made. The Aerio Cycle, featured at the Pan-American Exposition in Buffalo in 1901, looked like a giant seesaw with revolving wheels, fitted with cars, at each end. In 1968, the Swiss firm Intamin created a similar wheel, now called the Astrowheel, for the Marriott Corporation, which set it up at its Six Flags Astroworld in Houston. Eccentric (or sliding) wheels have, besides cars on the outer rim, others that slide on rails between the axle and rim as the wheel revolves. The most famous is probably Mickey's Fun Wheel at Disney's California Adventure

Park. This stands 160 feet tall and has 24 cars, 16 of which slide backward and forward while the others are fixed to the rim. The Singapore Flyer, which is 541 feet tall, was the largest when it opened in 2008, but the High Roller, opened in Las Vegas in 2013, topped it at 550 feet—the height of a 55-story building.[27]

The closely networked nature of design and production for technologies of play, as well as the immediate popularity of the ferris wheel, can be demonstrated by two American patents from the early twentieth century. In 1919, James J. Fairbanks, of Massachusetts, received a patent for "a new, original, and ornamental Design for a Toy Ferris Wheel." Two years later William E. Haskell, also of Massachusetts, applied for a patent for his design for "Ferris wheels for use on children's playgrounds."[28]

The roller coaster also had a long prehistory. Sometime during the seventeenth century, ice-covered downhill slides were constructed in Russia. Built of timber and covered with ice blocks, they had a set of stairs to the top of the slide, some 70 or 80 feet high, where riders boarded sleds for the 50-degree descent hundreds of feet from the top. By the early nineteenth century the rides, now called "Russian Mountains," had appeared in France and at about the same time (first either in Russia or France) wheels were attached to the sleds, which now ran on tracks.[29]

It was not until LaMarcus Adna Thompson built his Switchback Railway on Coney Island in 1884 that the roller coaster became popular in America. His 1885 patent for a "Roller Coasting Structure," having two parallel tracks "of undulating grades or planes," showed the two tracks as starting and ending at the same level and never attempting any great height. Its cars were said to travel at only about 6 mph.[30] Coney Island had its second roller coaster within months of Thompson's, and over the next century they evolved in much the same way as such other rides as the carousel and ferris wheel: they became bigger and faster, new designs were patented, and both individuals and family firms were involved over many years in their improvement, manufacture, and construction.

One significant difference, however, was that while these other rides were also designed to give pleasure and excitement, roller coasters were designed to provide thrills—that is, to scare their riders with a sense of *faux* danger. As historian Arwen Mohun has pointed out, the simple dichotomy of apparent danger and actual safety masked a complex calculus involving a masculine engineering culture, a heterosocial demography of consumers, and the need to design and operate rides that would be profitable. It is also important to acknowledge that while rides such as roller coasters were relatively safe for users, they were actually dangerous for the workers who inspected, repaired, and operated them.[31]

Variations appeared almost immediately. The latest amusement park thrill in 1902 was the Loop-the-Loop. "The car moves along by means of small rollers on wheels sliding inside a grooved rail," declared *The Cosmopolitan*, "and runs down a long incline with such speed that it makes a complete circle or turn in the air, the motion being so rapid that centrifugal force keeps the occupants from falling out, though they are strapped to their seats as an additional safeguard." But the Loop-the-Loop was just one of a number of such machines that had been devised to create at least the illusion of imminent disaster. Writing of Loop-the-Loops, Shoot-the-Chutes, roller coasters and similar devices, another observer admitted that "there is a strange fascination in this kind of motion [which] cannot be denied successfully. Fortunes are made by the proprietors of the opportunities to indulge in it, out of those who feel that they must shoot the chutes and loop the loops; while thousands of timid ones, who dare not try, are fascinated by the sight of the venturesome as they go careering and careening through the air." It was all, he wrote, "in these latter days of . . . marvelous progress," a deeply human attempt "to get ahead of time."[32]

The period of large roller coasters before the Great Depression was led by such firms as the Philadelphia Toboggan Company, which was founded in 1904, survived the doldrums of the 1930s and '40s, and took part in the revival of the rides after the opening of Disneyland in 1955. John Allen, its president after 1954, famously admitted that "You don't need a degree in engineering to design roller coasters, you need a degree in psychology. . . . A roller coaster is as theatrically contrived as a Broadway play."[33] One consultant for the firm was John Miller, sometimes called the "Thomas Edison of the roller coaster." He had been chief engineer for LaMarcus Thompson, had more than one hundred patents, and was especially important in inventing safety features which allowed ever faster speeds and sharper turns. Over the years the company was responsible for 147 wooden roller coasters, besides numerous carousels and other rides.[34]

The post–World War II revival of interest in roller coasters was marked by rides such as the Corkscrew at Knott's Berry Farm not far from Disneyland. The park had been accumulating attractions for many years, and in 1968 was finally enclosed, and admission charged. The 1975 Corkscrew took riders through spirals, creating a whole new experience of fear and excitement. Thirty years later the race for thrills was unabated, with Six Flags Great Adventure, in Jackson, New Jersey, opening its Kingda Ka. This new roller coaster launched riders from 0 to 128 mph in 3.5 seconds, taking the cars up 456 feet (the height of a 45-story building) before dropping them down the other side.[35]

In 1890, James Adair received a patent for an "Electrically-propelled Vehicle," which embodied the basic concept of what later became the popular Dodgem cars of innumerable amusement parks.[36] Adair's idea was to build "a platform or roadway or level surface of any desired dimension having a conducting-surface." In addition, "Above such surface at a suitable height and coextensive with it I erect a second conducting surface in the nature of a ceiling." His patent drawing shows a tricycle on the floor, connected to the ceiling by a trolley. When connected to an electrical supply, the vehicle could be driven about in any direction. Adair thought the device "particularly adapted for uses as a place of simple amusement," but believed it might also have practical applications. Apparently, the patented device was never actually built.

John Jacob Stock was closer to the mark with his 1920 patent for an "Amusement Device." His inspiration came from an unexpected direction. "In connection with amusement devices," he wrote, "it has been old to use what is known as a 'roulette wheel.' This device includes simply a stationary bowl in the center of which there is mounted a highly-polished disk capable of revolving, the circumference of which varies anywhere from ten to fifty feet. Occupants place themselves as nearly as possible in the center of this disk, and upon the same revolving they endeavor to hold themselves in place against the action of centrifugal force. When they are unable to do this they are thrown off and slide down into the stationary bowl."[37]

Stock seemed to think that having people sit in little vehicles "having universally rotatable wheels" was "an improvement" on the other form of the roulette wheel, but because the latter encouraged—almost required—people to frantically grab onto the closest other person to try to keep some balance, that judgment seems dubious. Stock admitted that "great amusement has been derived from the use of both," but believed his own improvement on the cars as superior. "My invention," he claimed, "contemplates the provision of a tub which shall be capable of movement in a forward or rearward direction, rotation in a clockwise or anti-clockwise direction, or pivoting around on a point. . . . [It] shall be entirely self-contained, even with regard to its supply of current." He imagined that it would carry ten storage batteries. He conceded that his tubs "may be provided with shock absorbing elements, as for instance, on the order of a pneumatic casing," but never admitted that the operator's "amusement" might actually be gained from ramming other tubs.[38]

Soon after Stock received his patent, another was granted to Max Stoehrer and his son Harold for an "Amusement Apparatus" which became the basis for their famous Dodgem cars. The Stoehrers deliberately equipped their device with "novel instrumentalities to render their manipulation and control difficult and uncertain by the occupant-operator." Indeed, they claimed that "in the hands of an unskilled operator," a "plurality of independently manipulated . . . cars" would "follow a promiscuous, irregular, and undefined path over the floor or other area, to not only produce various sensations during the travel of the vehicle but to collide with other cars as well as with portions of the platform provided for that purpose."[39]

Stoehrer, who was already manufacturing a ride called The Whip, reportedly frequented a nearby garage where he watched a young man with a stripped-down Ford speed around the facility and turn circles: it was fun to watch and, he thought, probably even more fun to drive. In 1920, before he had even filed his patent application, he tested his first Dodgem at Salisbury Beach, Massachusetts. It was a resounding success and in 1921 he and his partner, Ralph Pratt, who owned the beach park, formed the Stoehrer and Pratt Dodgem Corporation and began to manufacture more cars. The company would build a floor and roof for customers, then sell them the cars to fill it up. Subsequent patent applications for improvements were filed in 1920, 1922, and 1923, one of which was to make the cars easier to manufacture.[40] Over the years the Dodgem, with new models that looked more like real cars, became a standard in amusement parks. The company continued until 1961.[41]

For most of its history, the Dodgem's main competition was a bumper car designed and manufactured by the cousins Joseph and Robert (Ray) Lusse, who operated a machine shop in Philadelphia. A large part of the Lusse's business was making parts for roller coasters built by the Philadelphia Toboggan Company, but they had a broader interest. In 1911, they applied for a patent on what they called a "Roundabout," also called "human roulettes," which were "designed to afford amusement to children and others in their endeavor to stand or sit upon such rotating platform to maintain their position thereon, standing or sitting, during the rotation of the platform." Three years later they applied for a patent on an "Animal Mount or Support for Carousels."[42]

This intimacy with amusement rides meant that the Lusses were well aware of Stoehrer's success with Dodgem. In 1922, they applied for a patent for a "Vehicles Controlling Apparatus" to be applied to "motor driven vehicles, such as electric cars designed to carry one or more persons and commonly used in places of amusement such as parks, piers, or other places." Their new apparatus was "especially designed to permit the occupant or operator to easily govern both its speed and direction of movement." Whereas Stoehrer had made an advantage of the difficulty of steering the Dodgem, with its resulting "promiscuous" trajectory, the Lusses apparently realized that drivers not only wanted to bump other drivers, they very likely had specific drivers in mind.[43]

Over the years the Lusses took out eleven patents for modifications of what they called the Auto-Skooter, changing its "ornamental design," for example, and adding a new mechanism to allow it to turn more sharply. Over the years headlights were added, bodies were made of fiberglass, air-filled (instead of rubber) bumpers were added, and seat harnesses installed. The last Auto-Skooters made by the firm were sold in the 1990s, and since then nearly all such cars have been imported from Italy.[44] American developments continue, however: in the twenty-first century, bumper cars are being marketed for use on ice. Their manufacturer provides portable rinks that accommodate four cars and can be pulled to a suburban home for a children's party, but also recommends them to ice skating rinks that are seeking to bring "people back" to their businesses.[45]

All of these and similar machines were gathered together into the nation's growing number of amusement parks. An impetus to their construction was given by the 1893 Columbian Exposition held in Chicago. This great world's fair had a mile-long Midway Plaisance that led visitors to the central collection of white buildings which housed the artistic, cultural, and economic displays and exhibits. The whole site was popularly referred to as the White City. The Midway Plaisance was a mixture of crudely enthographic collections of "primitive" people from around the world and amusement shows and rides, including George Ferris's enormous "wheel." So popular was this area and its activities that other amusement parks also called "White City" sprang up in many cities around America and as far away as Australia.[46]

Paul Boyton's Water Chute opened in Chicago on the Fourth of July, 1894, inspired by the amusement enclosure at the Columbian Exposition and later claimed to be "America's first modern amusement park." The claim was based on the fact that previous parks had been based on a beach, gardens, or picnic groves. Boyton, on the contrary, built his park based solely on "mechanical attractions," especially the Shoot-the-Chutes ride. Encouraged by his initial success, Boyton built another such park on Coney Island, in New York, the next year. By 1908, Chicago had more amusements parks than any other American city, and others were added in later years. One of these was Joyland Park, owned and operated by African Americans during the 1920s in a suburb known as Bronzeville.[47]

Not surprisingly, by 1905 Chicago had a new "White City," spread over 14 acres and dominated by a high steel tower dramatically festooned with electric lights, creating an example of what historian David Nye has called "The Electrical Sublime."[48] Serviced by the South Side 'L' train, crowds arrived to enjoy the Racing Coaster ("The Greatest of All Thrills"), Chutes, "Flying Airplanes," a ferris wheel, a miniature railroad, incubators for premature babies, and a model of the Panama Canal. Over the years, airplane acrobatics were added and people could arrive via dirigible. Besides the technological attractions, there were burlesque shows and such vaudeville stars as the Australian swimmer and diver Annette Kellerman and the Wild West show of Buffalo Bill Cody.[49] After fires in 1911, 1925, and 1927, the park slowly died and finally, in 1939, it was condemned by the city and all but four buildings were destroyed.[50]

Cleveland, Ohio, was well served with amusement parks from the early 1890s, with the opening of Forest City Park and Washington Park. Both were on the Wilson Avenue trolley line. The former had a carousel but seems to have depended more on shows and concerts, along with bowling, dancing, and a shooting gallery to entertain visitors. The next, Euclid Beach Park, was opened in 1895 and became a fixture of Cleveland life. Initially the park relied on the traditional pleasures, with 1,700 feet of beach front along Lake Erie and 75 acres of woodland. Following a reorganization in 1901, admission was made free and only a nominal fee charged for an increasing number of individual rides, including a carousel, Thriller roller coaster, and "Flying Turns." Eventually more than one hundred rides and concessions were available. The park started a slow decline after World War II, then closed completely in 1969.[51]

A long-lasting but much-changed park opened west of Cleveland, near the city of Sandusky, in 1870. From the 1840s, sports fishermen had leased space on the peninsula, but in 1870 a German immigrant, who was a cabinet maker in Sandusky, built a beer garden and small dance floor there, along with some small bathhouses and "a couple of children's playground attractions." Visitors were transported to the site on his boat. This all seems to have lasted only one year, but in 1878 more bathhouses were put up, and in 1882, wooden walkways and picnic tables were added. Ten years later the peninsula's first roller coaster went up. Called the Switchback Railway, it operated by gravity and on occasion boys or horses had to pull the cars back to the top.

A Fred Ingersoll figure-eight toboggan roller coaster, the Racer, was built in 1902, but the park still had the reputation of being a beer garden. In 1906, a new "Amusement Circle" was installed, with a circle swing and a carousel. Beginning in 1908, with the Dip the Dip Scenic Railway, a series of roller coasters were built, often to be replaced after a few years with another. By 1969 the park had its Dodgem, Ceder Downs Racing Derby, San Francisco Earthquake Ride, Pirate Ride, Sky Ride, Mill Race, and Cadillac Cars. By 2010, Cedar Point claimed to have more rides than any other amusement park (75), but was widely known for one ride, offered in many forms.[52] From having no roller coasters in 1959, when the last one, the Cyclone, was pulled down, it had seventeen built between 1964 and 2007, including the Millennium Force "310-foot tall steel screamer" (built 2000) and the Top Thrill Dragster (built 2003), the cars of which reach a speed of 120 mph. Not surprisingly, it called itself the "Roller Coaster Capital of the World."[53]

In the nation's capital, the Washington Railway and Electric Company, which ran the city's streetcar system, bought land in nearby Montgomery County, Maryland, which had a small amusement park, including a carousel, already in operation. They renamed it Glen Echo Park, and added a dance hall in 1931, along with a roller coaster, new carousel, and bumper cars, as well as a swimming pool designed to accommodate 3,000 people at a time. Glen Echo Park was racially segregated, and as late as 1957 stubbornly declared that "Colored children are not welcome." Direct civil rights action finally succeeded in desegregating the park in 1961.[54] Across the Potomac River in Fairfax County, Virginia, the Great Falls Park was opened in 1906. The park was accessed by trolley and featured a Dentzel carousel, dismantled in 1952 but replaced the next year by a 1912 Stein & Goldstein carousel brought down from Rhode Island.[55]

The Carolina Power and Light Company developed Bloomsbury Park in 1912 at the end of its trolley line, just beyond the suburbs of Raleigh, North Carolina. A popular feature of the park was a Dentzel carousel dating from the turn of the century. In 1915, the company sold the carousel to the city, which moved it to Pullen Park, and soon after World War I, Bloomsbury Park was closed.[56] Playland Amusement Park, in Rye, New York, was opened in 1928 with a 1915 Mangels-Carmel "Grand Carousel" and a "Derby Racer" built by Prior & Church. The latter rotated at 25 mph, with its horses moving back and forth as well as up and down. A year later, Playland's famous "Dragon Coaster" was installed, and in 2004 it was still (including the 1928 wooden "Kiddie Coaster") one of five roller coasters in operation there. It is thought to be the "first totally planned amusement park in America," and was designed specifically to accommodate automobile travelers.[57]

On the West Coast amusement parks in the early twentieth century were usually placed on piers built out into the Pacific in frank imitation of England's famous Brighton Palace Pier. This structure was first built in 1823 to service ships crossing the English Channel to France. Damaged or destroyed several times by storms, it was rebuilt as an amusement destination in 1899, illuminated by 3,000 light bulbs, in themselves a technological drawing card. A collection of amusement machines were installed in 1905, making it a sort of "penny arcade," and by the early 1930s a Dodgem and Big Wheel had been added.[58]

The German-born Charles I. D. Looff, who had manufactured carousels in New York City, moved to Long Beach, California in 1910, built a new factory, and purchased land along "The Pike" to accommodate one of his large carousels. The Pike had begun in 1902 when "The Plunge," a large bathhouse, was opened on the beach at the terminus of the Pacific Electric Railway streetcar line from downtown Los Angeles to Long Beach. The Pike itself was a wooden boardwalk laid directly on the sand to connect The Plunge with the municipal pier, which was used for freight and passenger shipping. Gradually The Pike was widened, extended, and finally replaced by a concrete walkway lined with concessions and amusements. A large number of rides were installed, including Looff's carousel, Bisby's Spiral Airship (1902), and a "Figure 8 Roller Coaster" (1907) which rested on pilings and ran out over the ocean itself. In 1915 a new roller coaster, the Jack Rabbit Racer, was added. All told, at least twenty-five rides were in operation at one time or another.[59]

Looff built and operated amusement parks all along the nearby Southern California coastline, including at Ocean Park, Redondo Beach, Venice Beach, and Santa Monica. Indeed, between 1905 and 1912, there were seven amusement parks along the beach front of the three adjacent towns of Venice, Ocean Park, and Santa Monica.[60] A carousel built by Looff for the Santa Monica pleasure pier in 1911 burned down, to be replaced by another in 1916. To complement the carousel they also built a wooden roller coaster, the "Blue Streak Racer," as well as "The Whip" and an Aeroscope ride.[61] A carousel was built in 1911 for the Santa Cruz Beach Boardwalk, and in 1924, Looff's son Arthur built the Santa Cruz "Giant Dipper" wooden roller coaster. One-half mile long and decorated with 3,150 electric lights, its cars reached speeds of 25 mph.[62]

The transportation and real estate developer John D. Spreckels, who had studied chemistry and mechanical engineering in Germany, extended his San Diego Electric Railway (founded in 1892) to Mission Beach, where he had extensive real estate holdings. In 1925, he opened his Mission Beach Amusement Center, the outstanding features of which were "The Plunge," an indoor swimming pool, and a graceful wooden roller coaster, built by Prior and Church.[63]

The development of amusement parks in San Francisco demonstrates the continuing demand for such facilities as well as their important links to both streetcar transportation networks and the ocean beach. In 1868, hotel owner Robert Woodward opened his home and gardens, adjacent to his horse-car line at 14th and Mission streets, as a pleasure park available for a 25-cent admission fee. It seems to have been free of thrill rides but offered a more uplifting choice of gardens, museum, art gallery, and gymnasium.[64] In 1894, San Francisco's mayor, Michael de Young, opened the city's first "world's fair" in Golden Gate Park—the California Midwinter International Exposition. It featured a Midway Plaisance, with attractions from the previous year's Columbian Exposition in Chicago, as well as a "Scenic Railway" which was quite similar to a roller coaster.[65] Chutes Park opened in 1895 on Haight Street, featuring a 350-foot Shoot the Chutes, which was soon joined by a Scenic Railway, roller coaster, carousel, and other attractions. The Fulton Street Chutes, opened by Charles Ackerman, who was an attorney for several streetcar companies, had two of their lines passing its gates in 1902. It had a zoo as well as a theater and ferris wheel, and flourished until the great earthquake and fire of 1906. In 1907, a Coney Island Chutes was opened in the Fillmore District. After Ackerman died, his son moved his father's attractions to the Fillmore site and added a theater for movies and vaudeville, as well as new rides like the Devil's Slide and a carousel.[66]

By 1884, a steam railway was taking people to the northern beaches, and the first amusement ride at the site was constructed, the "Gravity Railroad" roller coaster, along with the Ocean Beach Pavilion for dancing and concerts. Soon three streetcar lines served the area, including the Sutro Railroad. Adolph Sutro, a German engineer who had made a fortune in the Comstock silver mines of Nevada, had salvaged the amusement rides from the 1894 Midsummer Fair and incorporated them into his new entertainment district on the northern beaches of the city, along with the Cliff House, Sutro Baths, and his scenic railroad. A line of concession stands lasted for two decades and was then dismantled.[67]

The Looff family had constructed an elaborate carousel for a site in San Francisco in 1904 but sent it instead to Luna Park in Seattle, Washington, because of the 1906 earthquake. Ironically it survived a fire at that park and was returned to San Francisco in 1914, where it was installed in the northern beaches site next to John Friedle's shooting gallery and baseball-throwing stand. Looff and Friedle became partners and by 1921 had ten rides, including Looff's Bob Sled Dipper, and because they also had a Shoot the Chutes, they called their park Chutes at The Beach. In 1922, they added Looff's Big Dipper roller coaster and eventually an Aeroplane Swing, the Whip, Dodge-Em, the Ship of Joy, Noah's Arc, a ferris wheel, and a hundred concessions.[68]

George and Leo Whitney arrived at the northern beaches in 1923 and by the next year were operating four shooting galleries, a quick-photo studio, and a souvenir shop. George, nicknamed "the Barnum of the Golden Gate," became general manager, renamed the park Whitney's Playland-at-the-Beach, and gradually bought out all the other concession holders. Eventually extending for three blocks along the Great Highway beachfront, it featured what one enthusiast called "toys for everyone": the 1912 carousel, the Big Dipper, Dodge-Em cars, a rocket ship ride, The Skyliner, and others. Equally popular was the Fun House, which featured "long wooden slides, a human turntable that spun and threw people off if they didn't hang on [to someone else!], distorting mirrors, and air jets that blew women's skirts up." Inviting people to enter was a giant mechanical woman, Laughing Sal. Her laugh, recorded on a 78 rpm phonograph record, was described as "a drunken yelping guffaw, an evil cackle, the uninhibited outburst of someone going out of her mind." Sal was commissioned from the Philadelphia Toboggan Company, which made amusement park devices, by the Old King Cole papier-mache company. Originally a "mechanical laughing department store Santa Claus," presumably with a far different recorded laugh, it was later fitted out with a "new head, female legs and breasts that jiggled on the end of springs."[69]

Playland thrived, during the Depression of the 1930s and World War II and on through the 1950s, as a working-class escape, but eventually became seedy and was finally sold and demolished in 1972. The Fun House was sold to the Santa Cruz Boardwalk, Laughing Sal went to the nearby Cliff House, and the 1912 carousel, initially sold to a shopping mall in Long Beach, was returned to the city in 1998 to become a part of the new Yerba Buena Gardens development.[70]

By the turn of the twentieth century, New York's Coney Island had become the iconic amusement park. That other modern technological miracle, the motion picture, had seized upon the park and made it the subject, or the object, of at least fifty-one films between 1897 and the 1960s. In some of these there were only fleeting glances of one or another ride, but others gave the park a more substantial role. The most notorious was Thomas Edison's 1903 film of the electrocution of the elephant Topsy, but there were also comedies such as "Fatty at Coney Island" (1917), starring the silent film comedian Fatty Arbuckle, with Buster Keaton in drag as his date.[71] By the mid-twentieth century screen depictions had become less common, but Coney Island was still such a powerful cultural icon that the Beat poet, Lawrence Ferlinghetti, could title one of his books of verse *A Coney Island of the Mind* (1958) without mentioning the amusement park even once. In a poem meant to be recited to a jazz accompaniment, however, he did complain that "I see where Walden Pond has been drained/to make an amusement park," suggesting that all of America had taken on the dynamic, tawdry, garish almost hallucinogenic marks of Coney Island—in fact had become one large Fun Factory.

It has been suggested that Coney Island's history can be divided into four periods: from 1829 to 1875 it was primarily a beach destination; from 1876 to 1896 it was a place where large hotels dominated, along with a midway; from 1897 to 1910 it evolved into an enclosed amusement park; and from 1911 to the present it experienced a period of growing crowds and eventual decline.[72] During the hotel era (in 1881), the Observation Tower from the dismantled Philadelphia Centennial Exposition was moved to the beach at Coney Island, providing, after riding up in a steam-powered elevator, a spectacular view from 300 feet above the ocean and city in the distance. Two double-decked steel piers were built, as well as a giant 175-high elephant which housed a hotel, restaurant, and shops.[73]

The era of mechanical rides began in the 1870s with the building of a carousel and in 1884 with Thompson's proto-roller coaster. The entrepreneur George Tilyou introduced a ferris wheel in 1894 and three years later opened Steeplechase, the first of three integrated amusement parks that would eventually define "Coney Island." The second, Luna Park, was opened in 1903, and in 1904 the third and last, Dreamland, was inaugurated, featuring one million electric lights. Over the next decades all three competed to build the best and most glamorous new rides—technological marvels that appeared with their speed and ability to disorient the conventional.[74]

Like other amusement parks, even the famous Coney Island suffered a gradual decline in the mid-twentieth century, partly as a result of the Depression and World War II. In 2010, however, the mayor of New York declared that "Coney Island is coming back, big time." Nineteen new rides, built by the Italian firm Altavilla Vincentina, were planned, as well as a "Scream Zone" featuring two roller coasters and a "human slingshot launching people more than 200 feet into the air." A spokesperson for the investors bragged that "we will have rides that will flip you, turn you, launch you, drop you, splash you and make the mayor want to lose his lunch."[75] As a reporter exclaimed, somewhat optimistically, "Watch out Disney."[76]

It was probably inevitable that the burgeoning number of amusement parks would seek to join forces to control and improve the political and economic environments in which they operated. Such parks were, after all, not just charming and exciting nodes of cultural expression, but private and (it was hoped) profit-making capitalist enterprises. The early twentieth century was a period in which associationism emerged and flourished in American society. Championed by such public figures as Secretary of Commerce (and later president) Herbert Hoover, the movement argued that individuals and enterprises with shared economic interests should band together into associations to present a coherent and forceful intervention into public affairs. It was seen, in short, as a way of addressing market problems without government initiatives.

As early as 1907, amusement park owners sought some way of presenting a common voice on public policy matters pertaining to their industry, and in 1917 they finally succeeded in forming the National Outdoor Showman's Association to subsume existing groups representing fairs, circuses, carnivals, or other more specialized enterprises. It soon became evident, however, that amusement park members were being saddled with most of the responsibility without the privilege of focusing exclusively on their own concerns, so in 1920, a National Association of Amusement Parks was formed to look after the interests of the approximately 1,500 such entities in the United States.

At the second annual conference of the new NAAP, three manufacturers of rides for the parks displayed their products to members. The following year that number grew to forty-five, and in 1925 the NAAP organized a committee "for manufacturers and distributors of amusement park equipment." Ten years later, following the logic of associationism, and no doubt in recognition of the fact that the interests of people who made rides and those who bought them were not necessarily always in harmony, the members of the committee broke away and formed their own American Recreational Equipment Association.[77]

Beyond the mid-twentieth century, the old Coney Island-style amusement parks were no longer the venue of choice for American families in search of a fun spot for a family outing. Increasingly, "theme" parks took their place, drawing heavily on the culture and experience of amusement parks (and sometimes little different from them). The distinct nature of theme parks was that the traditional rides and amenities of amusement parks were organized around a "theme" rather than simply being added willy-nilly as revenue and fashioned dictated. While certainly not strictly the "first" theme park, the one that established the rules and most successfully embodied the essence of the genre was indisputably Disneyland.

In 1993, *Sunset* magazine gave directions to what it called "Southern California's fun factories": for Knott's Berry Farm, "from State Highway 91, exit south on Beach Boulevard"; for Magic Mountain, "from I-5 exit on Magic Mountain Parkway"; and for Disneyland, "from Interstate 5 in Anaheim, take either the Harbor Boulevard or Katella Avenue exit."[78] In 1954, Walt Disney had begun, alongside the San Diego Freeway then under construction, to turn a 160-acre orange grove into an amusement park which he named Disneyland. It cost $11 million and opened on July 17 the following year. Disneyland was designed, according to one scholar, "for the values of long-distance travel, suburban lifestyle, family life, the major vacation excursion, and the new visual culture of telecommunications."[79]

Hugely successful from the beginning (1 million people visited during the first seven weeks), it inspired a second attraction, Disney World, which was opened near Orlando, Florida, in October 1971. Within a decade the two together drew more visitors than the nation's capital in Washington, DC. The Florida park was a whopping 27,000 acres, twice the size of the island of Manhattan.[80] Together, Disneyland and Disney World, along with Disney's later Florida attraction EPCOT (Experimental Prototype Community of Tomorrow), embraced twentieth-century American technology in three major ways. First, they depend upon transportation technologies, especially freeways and jet airplanes, to deliver visitors to the parks. Second, they make use of a host of sophisticated technologies to make the visitors' experience mirror Disney's precise plan for them. And third, they quite explicitly preached a gospel of happiness through technological progress.

These parks were above all "destinations." Michael Sorkin has insisted that "whatever its other meanings, the theme park [including Disney's] rhapsodizes the relationship between transportation and geography." As he points out, it is unnecessary for the average American family to go to Norway or Japan because those places can be reduced to "Vikings and Samurai, gravlax and sushi. It isn't that one hasn't traveled," he suggests, "—movement feeds the system, after all. It's that all travel is equal."[81] The trip to Disneyland is, in fact, a trip to many places at once, some real, most not, but all transformed by simulation. Once inside the parks, and out of the inevitable pedestrian "traffic jams," one rides through the Matterhorn, through the haunted house, or on Mr. Toad's Wild Ride. And people come by the millions. Orlando has more hotel rooms than Chicago, Los Angeles, or New York combined and is the "number-one motor tourism destination in the world."[82] Disneyland itself became a mecca not only for American families in their station wagons full of kids, but for visitors from overseas as well. On October 26, 1958, a Pan American World Airways Boeing 707 jet took off from New York for Paris, inaugurating an escalating frenzy of international jet airline travel. "We shrunk the world," exalted the pilot, Samuel Miller.[83] Any number of visitors from outside America found that a trip to Los Angeles, focusing on Disneyland, Las Vegas (a sort of adult theme park), and the Grand Canyon was the ideal American experience.

The technologies of the parks themselves were carefully thought out. As Sorkin has suggested, "one of the main effects of Disneyfication is the substitution of recreation for work, the production of leisure according to the routines of industry."[84] The senior vice president for engineering and construction of the parks was Rear Admiral Joseph (Can-Do Joe) W. Fowler, who had graduated from Annapolis in 1917 and completed a master's degree in naval architecture from MIT in 1921. He was in charge of engineering at both Disneyland and Disney World and completed both on schedule.[85] The irony was that, in an attempt to maintain the illusion of old-time virtues and simplicity, the "real" world of modern technology had to be hidden. Built on something like a platform, the park sits above a labyrinth of supply rooms and tunnels which allow both workers and garbage to be moved without being seen. In Disney's words, "I don't want the public to see the real world they live in while they're in the park."[86]

For his special effects, Disney's "imagineers" modified existing forms to create unique rides and experiences. In the Hall of Presidents, lifelike replicas nod and seem to agree with the bromides expressed by one of them. Swivel cars not only carried visitors through exhibits at a uniform and predetermined pace but also twisted them to face the correct direction for each presentation. The monorail became a model for others, particularly those used at airports. And finally, all the Disney parks, but especially EPCOT, preach a gospel of progressive technological determinism. As on observer put it, "the message everywhere is of technology ascendant, a new religion offering salvation, liberation and virtual shopping in a crisply pixellated sheen. Disney has an almost millenarian faith in technology."[87] The historian Michael Smith has analyzed the message of EPCOT and finds that, after the questioning of technology which peaked in America during the 1970s, Disney wanted to "revive the public's faith in progress, and in technology as the principal agent of that progress." He noted that for Disney, choice factors into technological change at the consumption end of the process, but seems absent from the beginnings of design and production. Further, any social or environmental disruption caused by technological change is accounted for by a "radical discontinuity between the past and future." The "humorless perfection of the future" is juxtaposed with the "whimsically flawed" though "steadily improving" past, with "the present poised uncomfortably in between." He concludes that "wrenched out of context in this way, events float free of causes and effects and history splinters into nostalgia," and "if decontextualized history is nostalgia, de-contextualized technology is magic."[88] The American past, like Tomorrowland, is the result not of our own choices, technological or other, but the work of some invisible and benevolent hand.

In an almost organic manner, the Disney empire has spread beyond America, even while the parks in the United States continue to grow and change. In 2012, some 119.1 million visitors crowded into Disney parks in California (two parks—Disneyland and the adjacent Disney California Adventure carved out of the original parking lot), Florida (eight parks), Paris (three parks, the first opened in 1992), Tokyo (two parks, the first opened in 1983), and Hong Kong (two parks, the first opened in 2005). A park in Shanghai is scheduled to be opened in 2016—a modern empire, apparently, upon which the sun never sets.

It was the genius of amusement parks, from children's fantasy lands through White Cities and theme parks, to transform pleasure gardens (and bare acres along a freeway interchange) into "Fun Factories" where a host of technologies operated to produce pleasure for the urban masses. Inventive talent combined with entrepreneurial daring and a shrewd sense of popular psychology and culture to produce sites of pleasure, excitement, and illusion. Crowds at these "pleasure places" were "engaged in the unalloyed pursuit of pleasure," according to Gary Cross and John Walton. These sites of "outdoor spectacle" invited "crowds to interact with each other and with the sites themselves, to participate actively as well as to gaze and listen, to move, mingle, compete for attention and put the self on show." The result, they suggest, was an "industrial saturnalia."[89]

It was all made possible not only by an improved and integrated transportation network, and a constantly growing and elaborate range of "thrill rides," but fundamentally by a cultural shift toward the acceptance of pleasure, something like that of the seventeenth and eighteenth centuries in England, which allowed the rise of Pleasure Gardens. As John Kasson has suggested, by the turn of the twentieth century, the keepers of middle-class respectability were being squeezed between the desires of the upper and working classes for amusements that were less earnest and "more vigorous, exuberant, daring, sensual, uninhibited, and irreverent." Formerly "marginal" activities became mainstream, and "the most striking expression of the changing character of American culture . . . [were] the new amusement parks."[90]

Arguably, much of the vibrancy and cultural risk of the classic turn of the century amusement park did not carry over into the new theme parks. A certain passivity characterizes the consumption of pleasure in a venue like Disneyland, whereas the earlier Playlands and Coney Islands had encouraged a degree of participation. Many of the earlier amusement park activities were exiled to the fringes of smaller, traveling shows, circuses, and carnivals, themselves devoid of any of the middle-class respectability that seemed to reign at theme parks. This shift merely underscores the social significance of amusement parks in shaping a modern industrial culture that had to make room for a large and demanding working class at the turn of the twentieth century.

Chapter Four

The Hobbyist

Making models makes model boys.   —CAPTAIN COMET

No gear, no hobby.   —COMMON SAYING

The April 1933 issue of *Popular Mechanics* featured what it called "The Boy's Own Room and Museum." On a page, separated by the expected detailed and dimensioned sketches of cabinet parts (which could easily be made in the home workshop), two scenes pictured a room for what must have been, for many families, the Ideal Boy. In the top sketch, our young hero sits reading in a window seat flanked by cupboards clearly marked "Nat History, Historical, Stamp Albums, Specimens, Relics, Scrap Books, Geology, Botany, and Misc." Atop the cabinets are model ships (sailing vessel and luxury liner) and a trophy cup. On the wall are pictures of a hunting scene and another ship, maps, a ship's steering wheel, and an airplane propeller. His neatly dressed friends are gathered at a table, looking through a microscope.

The bottom scene shows the other side of the room with bunk beds, a butterfly collection, sports equipment, a stuffed fish, and, on one chair the boy again, dressed and ready for the woods, in another chair, his father with a shotgun on his lap. The boy is petting his dog. This 1933 idealized space for boys embraced and illustrated the diverse reach and meaning of the world of hobbies.

But an earlier America had a much more ambivalent attitude toward the subject. The historian Steven M. Gelber has argued that, as industrialization sharpened the distinction between work and leisure, and the venues appropriate to each, "the fluidity of preindustrial time gave way to discrete blocks of commodified time that could be sold for work or withheld for leisure, which led guardians of public morals to fear that time spent not working would be time spent getting into trouble." Reformers worked on two fronts: first, trying to limit the availability of "inappropriate activities," and second, trying "to encourage benign pastimes. Hobbies have been among the most prominent of such socially approved activities."[1]

Amusement parks, such as Coney Island, appeared to be epitomes of leisure values, though in fact they were replicas of industrial America, which reinforced workplace values. And the bleeding of values from one site into its opposite ran the other way as well. As Gelber writes, "the ideology of the workplace infiltrated the home in the form of productive leisure." Furthermore, if modern workplaces were becoming more and more depersonalized, robbed of initiative and satisfaction, the new nineteenth-century hobbies could at the same time "condemn depersonalized factory and office work by compensating for its deficits while simultaneously replicating both the skills and values of the workplace," a process that Gelber calls "disguised affirmation."[2]

Before 1880, a hobby was often considered a dangerous fixation, as in the term "hobbyhorse," but after that period it rapidly was considered a productive use of leisure time, most often in the form of handicrafts or building collections, though it could also mean gardening or amateur participation in music, theater, sports, and so forth. Craft projects and objects collected shifted somewhat over the decades, but the practice remained essentially the same. For one thing, they were rigidly gendered. Sewing was a major craft activity for women, a genteel way to extend a feminine skill beyond any semblance of need. Victorian men avoided any manual activities at all until the Arts and Crafts movement of the late nineteenth century allowed room for what Gelber called a "domestic masculinity" which reached a crescendo in the do-it-yourself craze of the 1950s. By owning their own tools and producing something useful, both men and women tapped into the romantic myth of the sturdy artisan of preindustrial times. As Gelber put it, "male and female workers alike sought leisure that provided them with the meaning they felt missing on the job, but they did so through activity that was a better job than their job."[3]

Historian Rachel Maines places the emphasis elsewhere. "Once production shifts to industrial methods," she notes, "the leisure consumer is free to seek pleasure in the older handcraft technology." Indeed, she claims, "any technology that privileges the pleasures of production over the value and/or significance of the product can be said to be a hedonizing technology." As such they are "paths to pleasure."[4] To further complicate the question, it has been suggested that, espe-

cially for men, hobbies can be an escape from potentially difficult and chaotic in-
terpersonal relations and family responsibilities. A hobby can isolate, and provide
"a promise—in part symbolic, in part material—of mastery over the uncertain,
uncontrollable and potentially chaotic."[5]

Tools, in many ways, are also tests of gender conformity. In general, heavy
tools were reserved for men and lighter ones for women. This was an imprecise distinction, of course, but the boundary was patrolled assiduously: hammers
were for men and needles for women. Paint brushes represented something of a
liminal technology, though painting a watercolor or a chair was less controversial
for women than painting a whole room. These gender distinctions were passed
along intact to children, with boys being taught to hammer and saw while girls
were taught needlework. One difference was that mothers generally knew how to
sew but many middle-class fathers, at least, were unused to tools of any kind. Just
as women had servants to do their housework, if possible, men paid craftspeople
to take care of maintenance and repairs around the house. As late as the 1970s,
when one engineering school offered a course in hand tools to female first-year
students on the assumption that they had not been taught to use them, the male
students petitioned to have the same course, saying that they had not been taught
to use them, either.

One hobby that sometimes blurred the line between work and fun was printing
on a small printing press.[6] Between 1852 and 1875, at least twenty-eight presses
were patented for the use of boys, or at least amateurs. The clamshell press patented by C. G. Havens and F. C. Penfield in 1872, for example, was described as
being "a small press of cheap construction—such as are commonly known as
boys' or toy presses." By this time more than twenty firms were marketing such
presses, and more than fifty were at work in Washington, D.C., alone. A list of familiar American names of those who owned presses as boys includes the Wright
brothers, Thomas Edison, H. L. Mencken, Charles Scribner, John Wannamaker,
and Cornelius Vanderbilt. In 1869, General Ulysses S. Grant, then president of
the United States, ordered a Novelty Printing Press for his son. Three weeks later,
the White House sent a note to the maker saying that "Master Jesse is much
pleased with it all." The architect Frank Lloyd Wright, in his autobiography, reported that he and his friend Robbie had "a small printing-press with seven fonts
of De Vinne type, second-hand, . . . set up in the old barn at first, and later a quite
complete printing-office in the basement of the house." He called it "a real toy—
the press—for growing boys as well as for grown-up rich men."[7]

The presses were not simply scaled down professional machines, but modified devices that drew upon such office technologies as rubber stamps and copying presses. Originally conceived of as allowing small entrepreneurs, such as pharmacists and dressmakers, to avoid having to hire professional printers for their labels and such items, their advertising materials boldly proclaimed, "Every man his own printer!" or "Do your own printing." Such self-printing could save a shopkeeper money, but for others it could mean a whole new income stream. One advertisement counseled, "Boys. Don't be satisfied, and don't rest till you own a printing Press, and see for yourself the fun there is in it. Our word for it, you will never want for amusement or pocket money after you get a Press." Another advertised, "Business Men save expense and increase business by doing your own printing and advertising. For Boys delightful, *money-making* amusement." A thirteen-year-old boy from San Francisco wrote in 1876 that "I now have about sixty dollars that I have made with it. I go to school every day and print in the afternoon." H. L. Mencken was perhaps more typical. At the age of eight, in 1888, he was given a press and earned two dollars over the Christmas season. In later years he confessed, "So far as I can remember, my father was my only customer." Typically, it was fathers who gave their sons a press, although one customer wrote to the manufacturer to order a press, adding, "I design to make it [printing] amusing and instructional recreation for some young girls[,] nieces of mine." By the turn of the twenty-first century, of course, boys and girls who wanted to print could more easily do so using a computer.

For children who wanted to connect with the most modern of technologies, model airplanes were available from 1904, only months after the successful first flight by the Wright brothers. A magazine reported in 1913 that boys were making flying models of planes in such numbers that "a national model aeroplane club is forming from the many local clubs in existence." The boys, it was claimed, "while profiting in most ways by the experience of the real bird men and scientific model builders, work out their own planes from actual experience in flying machines." The article concluded that "the significance of this is evident, for with the coming of the new industry—as come it must—there will be a great demand for practical aerial engineers. And there is no surer foundation for success in this line of work than this early training of the boy [sometimes even in school shops] in the fundamentals of aviation."[8]

The evolution of model airplane construction can be divided into three periods. From their first appearance through World War I, modelers preferred "racers," simple designs often created by boys (seldom girls) themselves. Between the wars, racers were largely supplanted by scale models of actual aircraft, increasingly made from balsa, a wood from Ecuador that was not commercially available before the late 1920s. After World War II, newly available plastics were used to create scale models which required only punching out and gluing together. These last, in fact, largely marked the end of the model airplane as an important element in the technological play of American boys.[9]

During the first two decades of the twentieth century, a nascent model airplane manufacturing industry tried to interest boys in scale models that were replicas of such real airplanes as that of the Wright brothers. The financial interests of manufacturers were combined with an ideology of air-mindedness, requiring, as historian Aaron Alcorn has written, "models to be more than toys, and instead serve as tools for industrial training that promoted a distinct vision of modern boyhood—and, by extension, manhood—characterized chiefly by its engagement with technology."[10] Boys, on the other hand, proved more interested in building models that would really fly and adopted designs that resembled no actual planes but flew much more successfully than commercial replicas, thus imposing on the market their own preference for performance over appearance.

An early example of this style was the racer designed by Percy Winslow Pierce. Featuring a stick-like fuselage with two sets of wings (a longer at the rear and shorter near the nose) and a propeller at the rear, Pierce's model was so successful that in 1910, when he was only fifteen, he was manufacturing them in the attic workshop of his own home for sale to other boys. He then made a licensing arrangement with a local book publisher which marketed kits for making what they called the Percy Pierce Flyer, selling either the plans and instructions alone for fifteen cents, or the entire kit including materials for $1.15. The plane was so successful in flight that it sold, unchanged, for decades.

The growing popularity of the racers was both the cause and effect of the spread of model airplane clubs, many of which sponsored contests for their members. In 1909, Pierce made a flight of 152 feet, which was sufficient to make him a local hero; by 1912, the record for distance stood at a little over half a mile. The credit for this massive increase lay with the development of the "A-frame twin pusher," a model with two fuselages joined at the rear and powered by two propellers mounted forward. Apparently this was a British design but was widely adopted in the United States. In a contest in 1911, Cecil Peoli, then only sixteen, set a new record of 1,691 feet. He then licensed his A-frame design to the Ideal Aeroplane and Supply Co. which introduced, the next year, its Cecil Peoli Racer. It remained the most successful racer of the next two decades, while Peoli himself turned professional flyer and was killed in a crash in 1915. Other firms began to turn out kits for A-frame racers, making product differentiation a real problem. Meanwhile, popularity of the relatively inexpensive racers proved to be a block to the sale of more expensive (and therefore more profitable) items like Ideal's Wright Flyer, which sold for six dollars as a kit and twenty dollars pre-assembled. The desire of boys to design, assemble, and modify racers, even if they began with a kit, meant that they were able to remain producers even as they were becoming consumers as well, spending the "allowances" of twenty-five to fifty cents a week that were becoming common for middle-class children.

The tensions in the evolving hobby, and the industry growing up around it, were palpable. For example, the emphasis on performance in competitions demanded standards for parts and assembly that were difficult to maintain under the regime of what Alcorn has called the *bricolage* of "resourceful tinkerers."[11] A purchased propeller saved time and usually worked better, but even the search for bits of wood and wire to make one's own parts usually involved shopping, albeit at the lumber yard rather than the hobby shop. The tension between purchasing and improvising was real: too much of the former could undermine a boy's status as a hobbyist and define him as a mere consumer. Pressure from manufacturers to buy kits, especially scale models of "real" planes, pushed boys further into the market. The latter were evolving quickly, especially after the war and spurred on by research from the new National Advisory Committee for Aeronautics (NACA), and boys who continued to build only A-frame racers quickly lost touch with the aeronautical progress.

The second period of modeling gradually solidified after the war, in part as a result of the pressure from manufacturers and partly because of the availability of new materials. While a Percy Pierce Flyer, made largely of bamboo and silk, could stand as a typical product of the prewar years, by the 1930s boys were more likely to be cutting the parts for a replica Piper Cub from sheets or strips of balsa wood and covering the bodies as well as the wings with tissue paper. As Alcorn notes, "the boy who tried his hand at the De Havilland DH-4 [from the Ideal Aeroplane and Supply Co.] , . . . submitted himself to the logic of the kit, which for this model rationalized the process of procurement *and* production."[12]

The popularity of balsa wood kits with their rubber band-driven propellers allowed a younger cohort of boys to participate in the hobby, but another innovation of the 1930s, the small internal combustion engine, favored an older group of modelers, including some who had previously given up the hobby and were moving into young adulthood. It was also a period of intensified organization, with the founding of the Academy of Model Aeronautics and the Model Industry

Association. Magazines devote to the hobby, like the *Model Airplane News* and the *Model Aircraft Industry Coordinating Bulletin*, also appeared. Both the organizations and the periodicals were born out of the world of the hobby itself and catered to its needs.[13]

In the years immediately following World War II, the Comet Industries Corporation created Captain Comet, Jet Ace, to "introduce you to the fascinating world of aviation." The "you," of course, referred only to boys, as the back cover made explicit with the slogan "Model Building Builds Model Boys." In a period that valued family "togetherness" and when "juvenile delinquents" was a phrase increasingly heard and feared, the Comet company pointed out that "model building directs youthful energy into constructive channels; in the home, it becomes a unifying influence, as father and son work together in building faithful replicas of famous airplanes."[14] Though the hobby and its associated industry had changed radically, the ideology had not.

Another modern technology also attracted enthusiasts. As historian Kristen Haring has pointed out, "the hobbyists called 'hams' initially turned to radio for technical challenges and thrills." Furthermore, as "Geeks with an adventurous side . . . they were, in this sense, precursors to computer hackers," she concludes. The historian Susan Douglas calls them simply "audio outlaws."[15] Hams made their appearance in the first decade of the twentieth century, soon after Guglielmo Marconi initiated American broadcasting with news of the progress of the America's Cup races. Inexpensive crystal detectors, available from 1906 and still constructed from kits in the 1950s, made it relatively easy for large numbers of boys and young men to listen in to the rapidly growing radio telegraph traffic produced by both business firms and the federal government. When Lee De Forest began the era of radio telephony by broadcasting both music and the human voice in 1909, the nation's hams enthusiastically participated in that activity as well.

Unlike most corporations and government agencies, which sent point to point messages in Morse code, many amateurs began to transmit and receive broadcasts. This refusal to be limited by the prevailing notions of the way wireless should be used was part of why Douglas called hams "Audio Outlaws." With radio, and later with phonographs, amateurs imprinted their own oppositional values and practices on the growing industry. By 1912, the *New York Times* claimed that there were "several hundred thousand" amateur operators in the country, a scale of activity that was hardly affected by the passage by Congress in that year of a Radio Act aimed at stifling the activity. In the period 1915–1916, the Commerce Department licensed fewer than 200 commercial shore stations but also 8,489 amateur stations, a number that was supplemented by an estimated 150,000 illegal ones.[16]

Another and more pointed reason many hams were considered "outlaws" was that they did more than object to corporate and government domination of radio, they often took steps to disrupt it. One outstanding example was the intermittent warfare between amateurs and the U.S. Navy, which was a major user of available broadband. Some amateurs would send messages, pretending to be from official naval sources, to ships at sea, instructing them to turn back, change course, or, more specifically, to undertake fake missions like rescuing an imaginary ship in danger. More constructively, they gave congressional testimony in defense of civilian access to bandwidth reserved for the navy.[17]

Such activities helped give hams a popular aura of romance and daring. During the 1920s, at least twenty-nine titles were published in three book series celebrating the adventures of the "Radio Boys." Beginning in 1922, the earliest volumes often contained technical instructions, such as how to build a crystal set receiver, though they soon shifted to routine tales of boyish adventure. *Radio Boys Loyalty, or Bill Brown Listens In* (1922) was perhaps typical in both style and plot: the book opens with a college-bound friend exclaiming, "They've got a splendid broadcasting station at the Tech, Bill." Bill replies, "I know it; hence my general exuberance. And if we don't get at it once in a while, it'll be because we can't break in."[18]

The Radio Corporation of America (better known as RCA) was formed in 1919 to buy out the Marconi radio interests, and planned to continue emphasizing point to point Morse code messages. Amateurs, however, had already created a sort of broadcasting network and audience and therefore, as Douglas writes, "the radio trust had to reorient its manufacturing priorities, its corporate strategies, indeed, its entire way of thinking about the technology under its control." Within a decade it was RCA and not the amateurs who dominated the industry.[19]

Hams continued to thrive, however, their numbers expanding rapidly during the 1950s. By 2000 there were more than 680,000 in the country, 95–99 percent of whom were white men from the middle and upper socioeconomic classes, not inclined to identify ethnically, and more likely than average to have some college education and work in a technical field. In a period during which manufacturers simplified radios for the home "in order to attract female customers," ham radio outfits were powerful markers of masculine power and competence.[20]

Indeed, like other technically based hobbies, ham radio provided a homosocial activity that demanded a separate space in the home, beyond the assumed female control of normal domestic arrangements. It provided a technical definition of masculinity devoid of obvious sexuality, but at the same time thereby put pressure on social relationships. Whether his equipment was in a bedroom closet, the basement, the attic, or a shack in the backyard, the ham claimed a space that was his alone, forbidden to wives and daughters and available to sons only on the strictest terms. Under male control, these spaces could be messy and cluttered with equipment and did not have to be held to domestic standards of good housekeeping like the rest of the house. Importantly, time spent both on the air or making technical adjustments and improvements was time, by definition, during which the ham was not available to his wife and children. Added to this was the fact that the often considerable sums of money spent on ever better equipment was also not available to the family for other purposes. It was little wonder that "marital tensions had become a standard trope of hobby culture."[21] In this, of course, ham radio was hardly unique.

For radio hams who either didn't want to design their own systems or shop around for compatible components, there were, after 1954, popular Heath kits available that contained all the necessary parts and clear instructions on how to assemble them. Electronics, however, were not the first Heath products. Edward Bayard Heath was born in Brooklyn in 1888, and at the age of twenty-one he designed and built his first airplane. In 1911, he went to work in the motorcycle factory of aviation pioneer Glen Curtiss. The latter's aircraft factory was next door and there Heath built his second plane. He bought a Chicago-based aircraft company in 1912 and founded his own E.B. Heath Aerial Vehicle Co. In 1926, he began to sell his Heath Parasols, which were powered by Henderson DeLuxe motorcycle engines. The planes sold for $975, but could be purchased without the engine for $690. The components for a plane, minus the engine, could be had for as little as $199. And finally, for the technically skilled but financially challenged, the plans were available for only five dollars. Heath was killed in the crash of an experimental airplane in 1931.[22]

In 1935, Heath's bankrupt company was purchased by Howard Anthony, who switched to selling accessories for light planes. After World War II, Anthony realized that electronics was a newly attractive field and he bought large quantities of cheap, surplus wartime electronics equipment, intending to organize them into kits, unassembled, to enable individuals to make their own devices. Importantly, this passing of formerly military technology into civilian use was paralleled by the discharge of thousands of former military personnel who, during the war, had been trained to use and repair it.

Anthony's first offering was the kit to make the O1 oscilloscope, which sold for fifty dollars. At the time, commercial electronic devices were assembled and soldered together largely by hand, even among commercial manufacturers; hence this was a process that a consumer could follow at home, given the parts and simple, clear directions. The savings in cost could be significant, as could be the satisfaction of building it on your own. Over the years, Heath made available a whole range of equipment, including high-fidelity phonograph components and, beginning in the early 1970s, an FM radio tuner with digital synthesis, and a digital clock. In 1978, the firm brought out its H8 computer, not the first kit on the market but in some ways the best. In 1982 it produced the Hro-1 robot kit intended to teach the principles of industrial robotics.[23]

The "audio outlaws" that Susan Douglas identified among the hams were very much evident within and along with the body of Heathkit hobbyists. The audiophiles were not satisfied, for example, with the sound limitations of commercial records, phonographs, and radios. The human ear can hear up to 20,000 cps (cycles per second), whereas 78 rpm phonograph records could only deliver 7,500 cps, and although AM radios could handle 10,000 cps, they usually operated at half that. Technical advances that took place during World War II made better performance possible. The new 33 1/3 rpm LP (long playing) records could record as many as 12,000 cps, which was twice as good as the old 78 rpm recordings. The actual delivery of this improved sound, however, was handicapped by the manufacturers of commercial phonographs who were slow to bring their products into line with these new possibilities. Audiophiles, including a large number of discharged service men who had gained training in electronics during the war, began to build their own equipment.

In 1951, the *New Yorker* identified the hi-fi craze as the fastest growing hobby in the country, and in 1957, *Time* claimed that the nation was beset with "a new neurosis" which they called "audiophilia, or the excessive passion for hi-fi sound and equipment." That same year *Business Week* reported that the phonograph industry, which had "once looked down on hi-fi fans as mere fanatics," was rushing to meet the demand for higher standards. They were, of course, late off the mark. In 1951, the new magazine *High Fidelity* began publication and in its first year gained a circulation of 20,000, and within two more years something like one million Americans had invested in custom-built sets. It was a characteristic of many of these audiophiles that once having an advanced piece of equipment they spent additional time and money trying to gain even marginally improved performance.[24]

Like ham radio, hi-fi was a masculine enterprise. Phonographs in the 1920s and '30s were not particularly associated with men, but hi-fi, with its roots in hobbyists assembling from kits and wartime technologies and skills, was powerfully identified with the man of the house and, in the world of Hugh Hefner's Playboy culture, with the affluence and sexual success of the urban bachelor. While ham radio hobbyists were usually able to escape the domestic sphere by setting aside a closet or shack, hi-fi enthusiasts could only achieve a virtual escape since the equipment, including the speakers, were a part of the home furniture, and the pure sound, cranked up to a level that allowed the owner to get "inside the music," could be heard by the entire family. The unhappy wife, as with ham culture, became a common trope often found in the periodicals of the time. A 1955 article in *Harper's*, for example, bore the title "The high fidelity wife, or, a fate worse than deaf."[25] Being noisy was a widely understood way for boys and men to claim space, especially outside the home, but the technological enhancement of hi-fi equipment allowed for both the excuse of high culture and the claiming of space inside the home.

With patience and some aptitude, anyone could assemble a Heath kit FM radio, hi-fi phonograph, or other devices without knowing much if anything about electronics. The assembly manuals, however, usually included a short section on "Theory of Operation," and the company developed a relationship with electronics correspondence schools. The lessons learned, for some, were more important than the theory of electronics. Apple's Steve Jobs, who had assembled at least one kit, was quoted as saying: "It gave a tremendous level of self-confidence, that through exploration and learning one could understand seemingly very complex things in one's environment."[26]

Heath closed out its line of kits in 1992, allegedly both because leisure time was no longer so readily available, or at least not spent on hobbies, and because ready-made devices were so widely available and so inexpensive. One commentator lamented that "no more will fathers teach their sons how to solder at the kitchen table. (Heathkit builders were more than 95 percent male.) No more will boys pass the Heathkit catalogue around like contraband during science classes. And no more will proud Heathkit owners announce 'I built that,' when switching on the stereo." Senator Barry Goldwater, a former U.S. Air Force general and the 1964 Republican presidential candidate, was said to have flown his private plane to Heath headquarters twice a year to buy kits. The end of kits, he said, "leaves the amateur, like me, no place to turn." By the age of eighty-three he was said to have already assembled more than one hundred kits.[27]

Radios presented a possibility for improved performance, as had the hi-fi phonograph. The broadcasting industry was heavily invested in, and deeply committed to, AM radio. When Edward Howard Armstrong invented FM (frequency modulation) in the early 1930s, broadcasters tried hard to block its development. For example, David Sarnoff, the head of RCA, tried to block investment in FM and quietly tried to persuade the Federal Communications Commission (FCC) not to allocate any of the radio spectrum for FM use. Its ability to produce higher fidelity, however, and its subversive potential for undermining the influence of dominant corporate power and popular commercial culture, appealed to the audio outlaws.

By the early 1940s, Armstrong had developed a small network of FM stations, but the reallocation of its spectrum by the FCC in 1945 made the few existing FM receivers technically obsolete. Interest declined until the 1960s, when the increasingly crowded AM band made FM the only feasible possibility for start-up stations in large urban markets. Between 1960 and 1966 the number of FM receivers increased by a factor of five, and the next year some 40 million homes were equipped with either FM or AM/FM radio receivers.[28]

While many of the hi-fi tinkerers quickly built their FM receivers from kits, other enthusiasts purchased theirs, especially high-quality sets manufactured abroad. The FM revolution had less technical challenge than hi-fi phonographs, but shared the same counter-cultural appeal. FM listeners, like their hi-fi brothers, were predominantly young and male, but more interested in the programming than the equipment itself. Better educated and more affluent than the average American, the FM listener demanded more "cultural" programming, with classical music playing a dominant role. With the coming of a new "Youth Culture" in the 1960s, FM became the home of early rock and roll and folk music. Even though FM, like hi-fi, soon lost its edgy pioneering style as more established commercial interests scrambled to take advantage of the new medium, in the early twenty-first century FM retains something of its mystique of high culture (as National Public Radio stations attest) while AM remains the home of blatantly commercial popular culture.

It was never likely that the car would escape the attention of hobbyists. In the mid-1970s the writer John Jerome, who had abandoned New York for rural Vermont, was seized with a powerful urge to rebuild a derelict 1950 Dodge pickup. He called the project a "post-technological adventure," but also admitted from the start that "what the truck will really be is a . . . hobby. What a quaint word it is, echoing the 1930s in my head. It'll be my model of the Eiffel Tower built out of toothpicks . . . unscrewing and rescrewing nuts and bolts."[29] Jerome was right: his

A hot rod on the streets of Oakland, California, in 1973. Such modifications were at the same time a hobby, an expression of personal affirmation, and a work of art. Photograph by author.

commitment to working on his Dodge was one form of what was, in fact, a broad and longstanding American activity of restoring and often modifying trucks, cars, and motorcycles.

Early in the twentieth century the ubiquitous T Model Ford almost *required* the kind of modification that consumers were prepared to make. Afterparts manufacturers offered add-ons and replacements with which owners could modify their vehicles for camping trips or a wide range of farm work, most importantly providing stationary power to run a range of machinery from saws to washing machines. Additional comfort could be attained from buying and attaching complete new bodies available either in one piece or in kit form. The T Model was an attractive object of modification because it was cheap, parts were easily available, and in its basic simplicity left so much that could be done to it after it was purchased. Making modifications to the engine to produce more power was a common goal. Throughout the 1920s and '30s, and with the dashing Model A replacing the T in 1927, greater speed and better handling were the major goals of the home mechanics.

By the end of World War II, this concentration on function was joined by a concern with form. Some car hobbyists began to aim their modifications toward making their cars sleeker: running boards were removed, chrome decoration, the hallmark of the postwar Detroit car, was stripped off and the holes plugged, chrome around headlights was removed and the lights molded into the fender, rear license plates were often sunk into the trunk lid and covered with a glass pane, and often, the rear, front, or both ends were lowered. With this work done, it was said that the car was "leaded."[30] Other owners set out to simply "restore" the car to its original splendor. Any car more than twenty-five years old could be considered an "antique," though not all makes and models would have been considered "classics."

A prime example of form over function was the "lowrider." After World War II, young Chicanos in Southern California and New Mexico acquired usually older and cheaper stock cars, then turned them into what has been called "an Ethnic Banner." Referring to these hobbyists, the artist and professor of Chicano Studies, Judy Baca, asserts that "what they had was a style. They co-opted an American icon. . . . They took junk and created art."[31] The signature modification was to lower the chassis, either at the front end or at both ends, and the preferred way of driving was very slowly. Novel details included doors that folded down to make casino tables and elaborate hydraulic systems that caused the vehicle to rise up and fall again in a kind of dance move. Often the trunk was loaded with speakers that gave music, especially the bass notes, a unique sound. The cars were also typically spray painted with bright colors in scenes depicting various celebrations of La Raza, including Aztec symbols and Our Lady of Guadalupe. Alongside the cultural symbolism, numerous nude female bodies and jail bars reflected popular experience (or desire).

While young Chicano men were expressing cultural pride in modifying their cars to go low and slow, young Anglo men in the postwar era were going for speed. Hot-rodding began in Southern California in the late 1920s with a few enthusiasts who raced their machines on the dry lake beds of the region. The practice burgeoned after World War II and quickly gained a reputation as a dangerous sport. In one day of racing, twenty-seven hot-rodders were killed in accidents at Muroc Dry Lake in California. In cities like Los Angeles, drivers would sometimes block off a street, then hold a quick impromptu drag race, hoping to disperse before the police arrived. On one night Los Angeles County sheriffs arrested 187 participants. In 1947, the state of California passed a law making it unlawful to participate in or even witness a hot-rod race. Police crackdowns on racing were gradually supplemented with attempts to promote safer driving practices and to provide dedicated venues for supervised racing. The National Hot Rod Association was formed in 1951, in part to promote safety. By 1955, there were an estimated 2 million hot-rodders in the country, and 45,000 car clubs.[32]

The amount of time and money invested in their cars by these 2 million enthusiasts was enormous. High school football coaches complained that boys were quitting their teams and taking afternoon jobs "to make enough to support their cars and girls." "For the devotee," one study reported, "hot rod races are but a small part of the hobby. Almost without exception, the true hot rodder had built his own machine, tested it, and taken it back to the shop for more work. A hot rod is never completed."[33] The magazine *Hot Rod*, which first appeared in 1948, championed an ideology of "activity, involvement, enthusiasm, craftsmanship, learning by doing, experimental development, display, and creativity, all of these circulating around the motor car in unpaid time." Normal American males were assumed to be "attracted to mechanics, tinkering, competition, and the search for success," all of which could be found in the hobby of hot-rodding. But although much of this leisure activity partook of the cut-and-try technique, *Hot Rod* also insisted that the emphasis "was not simply on working with metal but on theoretical understanding, scientific knowledge, and designing skill." One might wonder how many hot-rodders met this standard, but it stood as the manly ideal. In 1952, a consulting psychologist wrote that the activity associated with hot rods was "creative, educative, competitive, constructive, and masculine all of which are desirable elements in furthering the best in the American way of life."[34]

Early in the twenty-first century a much smaller-scale but equally passionate car modification began to attract attention. In 2005, a San Francisco newspaper reported that a local electrical engineer and "committed environmentalist" had spent "several months and $3,000 tinkering with his car." The car was a Toyota Prius, which now had extra batteries and the capacity to charge them from an electrical outlet in his home. Another engineer, this one teaching at the University of California–Davis, had "built a plug in hybrid from the ground up" as early as 1972, the year before OPEC (the Organization of Petroleum Exporting Countries) created a national crisis by cutting off oil shipments to the United States. Over the years he built seven more, including modified Fords and Chevrolets. Toyota had originally frowned upon homemade modifications to its Prius (perhaps because of the danger represented by the very powerful batteries), but eventually dropped its opposition, admitting that the company might be able to learn from the hobbyists. "They're like the hot rodders of yesterday who did everything to soup up their cars," suggested a Toyota spokeswoman. "Maybe the hot rodders of tomorrow are the people who want to get in there and see what they can do about increasing economy."[35]

Recreational home repair and improvement was not a feature of middle-class life in nineteenth-century America. Middle-class men—even those who had retained some knowledge of tools despite their white-collar status, hired tradesmen to do necessary work on their homes, just as their wives hired cooks, cleaners, nannies, and general servants. Working-class men, with knowledge of and access to the necessary tools, were sometimes able to build their own homes, but such activity was sometimes frowned upon.[36] In his 1911 book *The Principles of Scientific Management*, Frederick Jackson Taylor publicized the case of "Schmidt," a pig iron handler at the Bethlehem Steel works. Schmidt, according to Taylor, "had succeeded in buying a small plot of ground, and . . . was engaged in putting up the walls of a little house for himself in the morning before starting to work and at night after leaving." Far from being pleased that Schmidt was using his leisure time productively, Taylor was displeased that he had obviously not worked hard enough on his day job.[37] Although DIY (do-it-yourself) was preeminently an activity of homeowners, Schmidt was obviously not a mere hobbyist.

In twentieth-century America, however, it is doubtful that any hobby came near to having the widespread popularity, and presumed ideological significance, of home repair and improvement: the Do-It-Yourself movement. Its rise at the turn of the century was coincident with the Arts and Crafts movement which flourished in both the United States and England, with its masculinist characteristics of solidity, honesty, simplicity, and artisanal skill. It ushered in what Steven Gelber calls "the emergence of Domestic Masculinity," with the notion that the middle-class home was not simply a shelter from the cutthroat capitalism of the marketplace, but also a venue for the exercise of masculine values, particularly in the obligation to maintain, repair, and improve the family house. As he explains the term, "domestic masculinity was practiced in areas that had been the purview of professional craftsmen, and therefore retained the aura of preindustrial vocational masculinity. The two concepts are complementary, but domestic masculinity acknowledges the creation of a male realm inside the house."[38] It was at the same time less specific and more inclusive than the attic, basement, garage, or closet workshop.

Home improvement was very much a product of home ownership, and between 1890 and 1930 the number of privately owned homes tripled, while a higher percentage were owned by skilled workers than by professional families. During the Depression of the 1930s, federal home loans made it easier to buy houses, and after World War II the coincidence of no-down-payment loans to veterans and vast new mass-produced suburban tracts led to a surge in home ownership. Between 1940 and 1990 the country's suburbs grew 519 percent, and once again it was not simply a middle-class phenomenon.[39]

Following decades of slowly increasing popularity, DIY developed after the war into a widespread activity that was both economic and cultural. There were many pieces that came together to make DIY so popular. The new availability of relatively cheap housing (much of it deliberately left unfinished) was a key factor, and so was the wide acquisition of technical skills developed by both men and women during the war. Long weekends and summer vacations had become more common as the union movement pushed successfully for better wages, hours, and conditions. It was frequently noted that experienced tradespeople were in short supply and expensive to hire. But the desire to take up hammer and paintbrush also came from an optimistic belief that a new, freer domestic lifestyle could be had not only as an end result of home improvement, but that it would grow out of the process itself: husband and wife working side by side shaping their own environment.

Coining the phrase "virtuous consumerism," the historian Elaine Tyler May found that "Americans responded [to the anxieties of the Cold War] with guarded optimism by making purchases that would strengthen their sense of security. Investing in one's own home, along with the trappings that would presumably enhance family life, was seen as the best way to plan for the future."[40] Besides the house itself, and the cornucopia of appliances to fill it, this often also meant adding an extra bedroom, building a deck, or making other improvements. The mutually reinforcing ideologies of Domestic Masculinity and Companionate Marriage worked to make "Virtuous Consumption" a practice for both women and men.

On occasion the urge to keep the family safe in a dangerous world took more specific form. The Office of Civil and Defense Mobilization distributed a pamphlet titled *The Family Fallout Shelter* and the 1960 film "Walt Builds a Family Fallout Shelter," which was made with the help of the National Concrete Masonry Association and televised nationally to show the Average American how easily a shelter could be made. The pipe-smoking Walt claims to have spent only "a few evenings on it and a couple of weekends." Following the plans and list of materials in the OCDM pamphlet, he produced a cement-block structure in a corner of his basement and fitted it out with bunk beds and shelving and stocked it with canned food. He even planned to paint its walls to please his wife.[41]

DIY itself, of course, required not only time and desire, but tools and materials as well. DIY enthusiasts needed the traditional tools of the building trades, of course: hammer, saw, screwdriver, pliers, and so forth. The increasing availability and reliability of factional electric motors early in the century also led to new power tools being available. In 1914, Samuel Duncan Black and Alonzo G. Decker, operators of a small machine shop in Baltimore, Maryland, applied for a patent (granted in 1917) for what they called an "Electrically-Driven Tool." It was in fact an electric drill with pistol grip and trigger. It was designed, they wrote, to "overcome certain disadvantages inherent in most," that is, it could be operated with one hand while the other could be used to hold it in position. In 1950, as the DIY movement was gaining momentum, Black & Decker manufactured their one-millionth ¼ inch Home Utility drill.[42] Attachments made it possible for the electric drill to also serve as a sander, grinder or polisher. One commentator claimed its versatility made it "a motorized shop you can hold in your hand."[43]

Even A. C. Gilbert, the toy maker who had invented the Erector set, tried his hand with the electric drill. In 1929, he applied for a patent for an "Electric Drill or Similar Tool," which was granted in 1933. His invention, he claimed, "related to a relatively light, portable electric tool which may be supported in the hand of an operator, and which, while efficient in use, is of relatively simple form and may be cheaply manufactured." His drill looked something like Black & Decker's, though with a "novel switch arrangement" rather than a trigger.[44]

Husky Wrench was founded in Milwaukee, Wisconsin in 1924 by the Czech immigrant, Sigmund Mandl. The firm made hand tools, pneumatic tools, and tool storage products. Then, like so many small manufacturing firms, Husky went through a series of buyouts and mergers until it became a part of the Stanley Works in 1986. In 1992, Stanley began to market its Husky Tools solely through Home Depot.[45]

By the beginning of the twenty-first century the drills, hammers, and wrenches (and other so-called dumb tools) of both the professional and the hobbyist had been joined by a number of "smart" tools that relied on chips, lasers, and LCDs. Stanley Works was marketing a digital tape measure (with optical sensor and display screen) as well as a digital stud finder which could look through walls to find not only studs but pipes and electrical wiring as well. One commentator, who tried out some of them at home, remarked that "gadget fans will add one

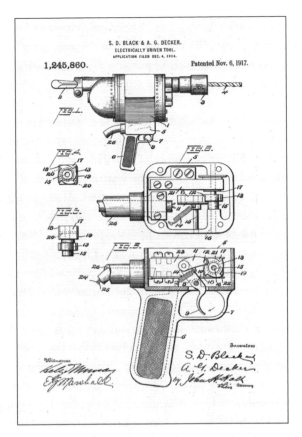

A century after its invention, the electric drill is still
a widely used tool of the DIY homeowner. Drawing
for patent No. 1,245,860 issued November 6, 1917
to S. D. Black and A. G. Decker for an "Electrically
Driven Tool."

or more of these smart tools to their toolboxes just because they are gadgets,
and therefore cool. Tool fans will buy them because they love tools." The editor
of a tool magazine for professionals, however, warned that part of the artisan
"tradition is learning how to use the tool, and knowing each tool's quirks. There
is something lost in the relationship between craftsman and tool when the tool
becomes smarter than the user."[46]

New building materials as well as tools helped make work easier for the home
improver. Plywood was increasingly used by builders during and after World
War II, and the introduction of new, smaller dimensions increased its usefulness

to home improvers. New kinds of gypsum board made it easier for DIYers to put up drywall, and in 1953, Reynolds Metals Company began selling "Do-It-Yourself Aluminum" which could be cut with a handsaw. Plumbing had been largely beyond the skill of amateurs until the 1960s, when PVC (plastic polyvinyl chloride) became available.[47]

Painting and decorating, usually considered the housewife's contribution to home improvement, became much easier with the introduction of water-based resin emulsion paints. Historian Carolyn Goldstein has attributed the development of these paints during the 1940s to the wartime shortage of lead and oil as well as to a desire to circumvent professional painters. After 1949 latex was used as the main binder. Such paints could be thinned with water, and water rather than turpentine was used for cleaning up. Sherwin-Williams' Kem-Tone, one of the early popular brands, advertised, "No muss, no fuss, no bother." A roller for applying paint had been patented as early as 1869, but it was not until Sherwin-Williams introduced its Kem-Decorator Roller-Koater that rollers began to replace brushes.[48]

Before World War II hardware stores and lumber yards changed only slowly to meet the needs of the "amateur" builder. Like the old-fashioned grocery store that was still common, they sold customers small quantities of goods taken from bulk containers: in the case of groceries, a wedge of cheese, perhaps cut from a large wheel by the grocer, or from a hardware store a pound of eight-penny nails taken from a keg. Hardware such as hinges or doorknobs had long been available through catalogs, and tools could be purchased at large chain stores such as Montgomery Ward or Sears. Then, following the supermarket model after the war, both hardware and grocery stores became increasingly self-service, with many items pre-measured and packaged.

The home improvement centers that began to spring up in the suburbs, catering to homeowners as well as contractors, were a quantum leap beyond the traditional hardware store. . Some, like Hechinger's in Washington, DC, stocked as many as forty thousand items.[49] When Home Depot opened its first two stores in Atlanta, Georgia in 1979, they immediately raised the bar for the mega-retailing of hardware and building supplies. Each store was a virtual warehouse, with wide aisles of stacked shelving. Open at hours that were convenient for home improvers, and featuring relatively low prices based partly on large volume, the stores drew customers away from smaller and more traditional retail outlets. By 1997, there were more than five hundred Home Depots in the United States, averaging 105,000 square feet in size. Some were even larger, such as the one in Union,

New Jersey, measuring 225,000 square feet.[50] By the twenty-first century, even such stores as Home Depot were courting the female builder with special classes in the use of tools and materials, or on particular projects such as bathroom or patio renovation.

As often happens, the triumph of such warehouse stores was followed by the creation of more boutique establishments catering to smaller markets. TechShop opened its first store in Menlo Park, California, in 2006. Other stores followed in Portland, Oregon and Raleigh-Durham, North Carolina, and in San Francisco in 2011. Jim Newton, who opened the first store, called himself "an active member of the maker movement," and claimed to have done it because he "needed space for all of my projects that were overflowing out of my garage."[51] Rather than customers buying tools, the TechShop catered to people who became members in order to use them. The San Francisco location, for example, had a half million dollars worth of tools. Tools on the ground floor included the "hard arts," e.g., milling machines, welders, and so forth. Upstairs were the "soft arts," including industrial quilters, vinyl cutters, a laser cutter, and twenty CAD (computer aided design) workstations. Classes were held in the use of the different technologies, and the membership, which paid a monthly fee, was drawn from "the ranks of hobbyists, artists, students and entrepreneurs."

During these same years, DIY in the kitchen seemed to have declined. Historian Rachel Maines has pointed out that in 2006, only slightly more than half (53 percent) of meals eaten in the nation's homes were actually made there, and of these "home-prepared meals," many were either sandwiches or bowls of cereal. Nearly half of food budgets were spent in restaurants, on takeout foods, or on already prepared foods bought at grocery stores. Her conclusion is unsparing: "clearly, most of those who still cook from scratch are doing so because they like to do it, and are investing in the tools, materials, and workspaces associated with culinary pursuits as a hobby."[52]

At some point in the late twentieth century, the family kitchen increasingly became the scene of cooking as a new hobby for men—yet another stage for the performance of Domestic Masculinity. Men had long been dominant as chefs in restaurants, and were usually designated to barbecue on the outdoor grill, but cooking inside the home had always been a chore undertaken by the housewife as a part of her culturally prescribed role as homemaker and family caregiver. By the early twenty-first century, however, an estimated 27 percent of American men were said to be the "primary food handlers for their families." It was interpreted as another example of "the yuppification of what we eat," and was accompanied by a rapid expansion of the market for kitchen tools. In 2002, a spokesperson for the Cookware Manufacturers Association was quoted as saying that "Growing male interest in cooking is one of the bright spots in the kitchen retail market." He went on to claim that "Men tend to have no problem buying a special pan for paella, if the recipe calls for it, whereas women will make do with a regular skillet or pan." "It's only since men have been cooking," commented one sociologist, "that you can justify the $275 knife."[53] The same might be said for the stick blender, the food processor, the juicer, and the authentic Moroccan tagine. Large, heavy restaurant-type gas stoves and massive multidoor refrigerators are expensive but are efficient and reassuring aids to serious cooking. In the early twenty-first century coffee machines, both freestanding and built-in, that used aluminum pods of grounds became common. The pods for the high-end Nespresso machines could be purchased in sleek, chic, minimalist boutiques in city centers. The masculine machine aesthetics of the shops seemed to promise that every man could be his own barista.[54]

Another observer noted, however, that "Cooking involves tools, dangerous, complex gadgets. Men who cook have well-stocked kitchen drawers. Every step involved in producing a meal requires a different piece of equipment, all of which needs to be washed up." Cleaning the pots and pans lacks the aesthetic and creative appeal of cooking, garners much less praise and admiration, but does not always have to be done by the person responsible for the preparation of the meal. The movement of men from the masculine space of the radio shack or basement/attic/garage workshop to colonize the formerly feminine space of the kitchen was no doubt in some way an important feminist victory, but only a partial one. As the previous observer also noted, for all the wives who could brag that their husbands "did all the cooking," she had never met one who could claim that he did all the cleaning of the bathroom or all the ironing.[55]

At the turn of this century the writer Novella Carpenter and her partner moved into a house in what was considered a bad section of Oakland, California. Wanting to eat well on a very limited budget, she set about turning her apartment, backyard, and the adjacent vacant lot into an urban farm. It began with an assortment of meat birds ordered online, and when they were delivered she immediately needed a brooder. Each new animal or vegetable seemed to require another tool or device. When they started keeping bees they needed a honey extractor. This realization reminded her of an old saying: "no gear, no hobby." Bees were followed by other acquisitions, including eventually a very large hog that she had slaughtered and from which she made all manner of processed meats like bacon and sausages. "The longer we lived in Oakland," she wrote, "the more garden related gear we seemed to accumulate."[56]

If one intended to modify one's car, one obviously needed such tools as a set of wrenches, and much more. If one wanted to become a serious bicyclist one obviously need a bicycle, but probably also lights, a tire pump, a water bottle, and perhaps even an odometer or GPS. To some extent the technology that accumulates around a hobby is simply necessary, or at least useful. One suspects, however, that for at least some enthusiasts it is the gear itself that is the desired goal—more end than means.

Chapter Five

Games and Sports

I, Frederick W. Taylor . . . have invented a new And useful Improvement in
Tennis Rackets.   —PATENT SPECIFICATION (1886)

The "world's first intelligent shoe" has a sensor that measures the
compression of the heel on impact with the ground.
                         —DESCRIPTION OF ADIDAS 1 JOGGERS (2005)

The comedian George Burns remembered that in the early twentieth century,
"our playground was the middle of Rivington Street" in New York City. "We
only played games that needed very little equipment, games like kick-the-can,
hopscotch, hide-and-go-seek, follow-the-leader. When we played baseball we
used a broom handle and a rubber ball. A manhole cover was home plate, a fire-
hydrant was first base, second base was a lamp post, and Mr. Gitlez, who used
to bring a kitchen chair down to sit and watch us play, was third base. One time
I slid into Mr. Giletz; he caught the ball and tagged me out."[1] With little money
to spend on commercially available play equipment, even bats and balls, urban
children became adept at making do with found resources. The greatest of these,
of course, was the cityscape itself, and, like the adult followers of the extreme
sport parkour later in the century, they adapted the built environment to their
own needs. Fire escapes, doorways, stoops, alleys, sidewalks, curbs, fire hydrants,
storm drains, utility poles, manhole covers, and even the pushcarts, parked delivery wagons, and adult pedestrians were transformed by the bricolage of play.

From pickup basketball games, through Little League teams, to big-time professional football, sports share certain characteristics: first they have rules, whether
well-enforced or varied to meet local and immediate contingencies; second, each
has at least an imagined venue—a playing field or court that helps contain and
define the play; and third, they make use of standardized equipment even though

sometimes it is not all available on all occasions. When George Burns and his friends played baseball in the streets of New York, nothing, except perhaps the ball, was regulation, but no one could doubt the game that was being played, or the enjoyment of the players. The importance of the "tools" with which the game is played is suggested by the threat that one of the players could always "take his ball and go home," thus ending the play for all and demonstrating, once again, the power that comes from owning the means of production.

An important part of any sport, presumably, is to exercise, improve, and demonstrate the prowess of the human body: to achieve the maximum possible speed, strength, or other attribute. The role of sport to celebrate and improve physical and mental achievement, however, has long been threatened by a parallel attempt to augment the human with the technological, to set a new record not by being stronger but by having an improved bat, or racquet, or pair of shoes. The convergence of inventors and manufacturers who want athletes to buy and use their products rather than a competitor's, and the desire of the athlete to set a new personal best, has meant a steady improvement in the tools of sport. This in turn has fed an ongoing debate to decide at what point a technical improvement substitutes for, rather than enhances, a personal performance.

Like the pleasure gardens of the eighteenth century, colonial American games and sports were massively altered and expanded by the Industrial Revolution in the nineteenth. Sports teams, events, and venues, for example, became tied together by a web of transportation and communications networks. Work on America's first passenger railroad, the Baltimore and Ohio, began in 1828. It was not finished for some years, but by 1850, the nation had something like 9,000 miles of rail lines. Ten years later there were 30,000 miles of track and four major trunk-lines joined cities on the northeastern coast to those of the Old Northwest. By 1890, railroads wove their way along 166,703 miles, tying together many of even the smaller cities of the country. These became the arteries of urban sports rivalry. Teams and crowds of fans traveled to games, just as horses and "sporting" crowds had moved by steamboat for races, and similar crowds traveled to converge on widely publicized prize fights in the antebellum years. As a strikingly urban phenomenon, baseball clubs seemed to spring up when rail connections became available. A National Association of Baseball Players was organized in 1857. The year before, Chicago got its first baseball team, organized just two years after the city gained a rail connection with Baltimore.

After the Civil War baseball expanded rapidly, coalescing at first in regional and then in wider associations. Teams from Detroit, Milwaukee, Dubuque, and

Chicago all attended a tournament in Rockford, Illinois, in 1866. In 1869, the Cincinnati Red Stockings made an exhibition tour from Maine to California, using the transcontinental railroad completed only that year. The organization of the National League took place in 1876, and ten years later the Michigan Central railroad bragged that "the cities that have representative clubs contesting the championship pennant this year are—Chicago, Boston, New York, Washington, Kansas City, Detroit, St. Louis and Philadelphia. All of these cities are joined together by the Michigan Central Railroad."[2] Even railroads were not enough to make the National League truly national, nor allow the American League to cover all of America. It was only with fast commercial air travel after World War II that the West Coast got major league teams (the Giants moving to San Francisco and the Dodgers to Los Angeles, for example). Until then the AAA Pacific Coast League provided the highest level of play throughout Washington, Oregon, and California.

College sports also benefited from the possibility of railroad travel. Although Princeton and Rutgers universities are both in New Jersey, when they played their first football game in 1869, students from the former traveled to Rutgers by train. Harvard University's baseball team made a grand exhibition tour by railroad in 1870, playing games against both amateur and professional clubs in New Haven, Troy, Utica, Syracuse, Oswego (Canada), Buffalo, Cleveland, Cincinnati, Louisville, Chicago, Milwaukee, Indianapolis, Washington, Baltimore, Philadelphia, New York, and Brooklyn.[3] In 1888–1889, Albert Spalding, the star pitcher and leading sporting goods manufacturer, took a group of major league baseball players on a tour, traveling through the American West, Hawaii, New Zealand, Australia, Ceylon, Egypt, Italy, France, and England, then across the Atlantic to New York, Philadelphia, and Chicago.[4]

Closely tied to railroads was the telegraph, with lines running along the tracks providing a nervous system for the railroads themselves. They also, of course, carried commercial and personal messages, as well as news. The telegraph network spread rapidly after the first message was sent in 1844 and San Francisco was connected in 1861. As the cost of messages decreased, the use of the medium to carry sports news increased. A prize fight had been reported in 1849 and the expanding "penny press" proved eager to carry the news of such events. Box scores, betting odds, and other sports-related news was printed in the large number of metropolitan dailies, but sometimes for well-publicized events, crowds would gather for results at Western Union offices and such male homosocial sites as poolrooms and saloons, which had wire connections. H. L. Mencken recalled that in Baltimore in the 1880s, upscale saloons would hire telegraph operators who would take reports and write scores on a blackboard for the patrons.[5]

Combined with the telegraphic transmission of news, technological improvements in printing were put at the service of sports reporting. The Napier double-cylinder press, imported from England by 1830 and further developed by the American firm R. Hoe and Company, greatly facilitated the rise of cheap tabloids such as the New York *Sun*. First simply reporting the results of horse races, prize fights, and even foot races, sporting news became an important part of the "new journalism" of such noted publishers as William Randolph Hearst and Joseph Pulitzer in the 1880s. By that time specialized sports writers, pioneering fast writing in a breezy style, made sports pages a regular part of big city newspapers. In addition, newspapers began to issue sporting almanacs listing upcoming events, records in various sports, and sporting news. The *Spalding Guide* and *Spalding Library* of books, both produced by the sporting goods manufacturer A. G. Spalding & Brothers, highlight the way in which publishing and the growing manufacture of sports equipment helped both create and cater to a mass market of sports fans and players, producing standardized equipment, keeping records, and setting standards, rules, and expectations in many fields of play.[6] Closely tied to the growth of great cities and tied together with sprawling railroad and telegraph networks was the rise of a manufacturing industry firmly based on such technological developments as steam power and new modes of production, and carefully studied by historian Stephen Hardy.

As early as 1834, William Clarke's *Boy's Own Book* gave boys instructions for making a baseball by hand: how to cut India rubber, wind it around a cork core, wrap that in wool yarn, and sew a leather cover around the whole thing. By 1887, A. G. Spalding was mass-producing balls and selling them for as little as five cents each.[7] With the establishment of a sporting goods manufacturing sector also came sports sponsorship and the rise of "official" (that is, standardized) equipment and rules.

Although the early nineteenth century saw the establishment of a few journals, such as the *American Turf Register*, which discussed sports and encouraged their practice and spread, Americans still played with equipment that was either imported, homemade, or at least locally manufactured by artisans. The two decades from the start of the Civil War witnessed the rise of a recognizable industry: the 1870 census, for example, was the first to list "baseball goods" as a separate category of manufacture. The "new sports" of the time—baseball, croquet, football, tennis, and bicycling—were widely seen as "exotic and frivolous," hardly safe areas for investment. Manufacturers gradually moved over from producing goods for the older sports, and from non-sports products, and sometimes well-known players decided to move from the playing field to the factory.[8]

The most prominent player to turn to selling equipment was Albert G. Spalding, a pitcher for the Boston Red Stockings and, later, the Chicago White Stockings. A highly successful player (career ERA of 2.14 and batting average of .313), Spalding opened a retail sporting goods store in 1876, the same year he won forty-seven games for the White Stockings. Two years later he and his family partners bought a firm that manufactured bats, skates, fishing gear, and croquet equipment and became the first retailers to manufacture and sell their own goods. Spalding was the first star to wear a leather glove on his catching hand, a move that probably increased sales of the glove. The census of 1880 was able to count eighty-six manufacturing firms in the new category of "sporting goods."[9]

In the 1880s, newer sports such as football and tennis moved from a dependence on imports and local artisanal products to large-scale manufacturing. The Rawlings Brothers opened what they called a "full-line emporium" in St. Louis in 1888. Philip Goldsmith of Covington, Kentucky, manufactured toys but eventually realized that his best-selling items were inexpensive baseballs, and shifted from toys to sports equipment. The Al Reach plant in Philadelphia was a leader in quality and consistency of baseballs and sought, through "rationalized production, greater volume, and wider price ranges," to control Reach's market share. High-grade balls remained a labor-intensive product, which required more than a thousand workers in the factory preparing balls that then had their leather covers stitched by hundreds of women working under sweatshop conditions in their own homes. Prices ranged from $1.50 for the "Association" balls to five cents for the balls used by kids in dusty fields and side streets all over the country. Reach produced a leather glove for twenty-five cents and a top-line buckskin model for only $3.00.[10]

The proliferation of sporting goods firms led some of them to try to stand out from the crowd by publishing guide books. In 1876, Spalding attained exclusive rights to the "official" National League book, but went further to create *Spalding's Official Baseball Guide*, which for ten cents provided "not only the league rules and constitution but also records, descriptions of the past season's play, instructions, and history." Readers, of course, were also provided with information about Spalding equipment. Spalding also made an advertising coup by agreeing to pay the National League a dollar a dozen for the balls it provided free, giving the company the opportunity to advertise them as the "official" League balls. Other suppliers made similar deals with other leagues.[11]

The increase in the number of manufacturers counted by the 1900 census, 144 firms, up 67 percent over 1880, was clearly tied to the increased participation in sports. Some equipment manufacturers worked hard to create a demand for their products, none more so than Spalding. His firm designed the sports stadium, organized the competition, and provided the equipment for the 1904 St. Louis Games. He promoted sports of all kinds, giving trophies to schools around the country to be presented to various league winners. Although perhaps best known for baseball equipment, Spalding was the first American manufacturer to make golf clubs and balls (1894 and 1896 respectively). Again, in 1899 he sponsored a national tour by one of the game's premier players to promote interest around the country. The Boston baseball player and equipment merchant George Wright was also a promoter of golf, importing his first clubs from England in 1890 and helping to turn a game, almost unknown in America, into the craze of the country club set. He gained permission from the Boston Parks Commissioners to play in Franklin Park for the first time, and, partly through his efforts, twenty-nine courses had been built around Boston by 1899. Roller skating, which had had a brief flurry of popularity in the 1860s, was revived in Boston in the 1890s when J. M. Raymond, head of the Raymond Skate Company (founded in 1884) opened his Olympian Skating Club. The bicycle manufacturer Albert Pope recalled that, in the 1870s, people had to be educated "to the advantage of this invigorating sport" of cycling. Because they had to create as well as serve a market for their equipment, "sporting goods firms," as Hardy insists, "became the front line of promoters and educators during the great sports surge of the late nineteenth century."[12]

Equipment manufacturers not only stimulated and supplied a whole list of growing sports, it also had a large role in shaping their play. Leagues and associations rose up to regularize competition, often calling on manufacturers to supply "regulation" equipment. But innovations sometimes tended to disrupt recently standardized play. In football, boxing, and ice hockey, new padding and protective designs had the result of making the sports more violent: in ice hockey, for example, encouraging "the clashing, hard-checking 'Ottawa' style of play." New rubber-thread-wound golf balls, developed by Coburn Haskell, "turned more golfers into long hitters and caused wholesale changes in course design." The rise of "ready-made" equipment, rules, and styles of play was often associated with

In 1912, the Cincinnati Reds moved into their new ballpark, Redland Field, renamed Crosley Field in 1934 and used until the team moved to Riverside Stadium in 1970. The relatively new reinforced concrete and steel structures were longer lasting and larger than their nineteenth-century wooden precursors. Courtesy of the Lake County (IL) Discovery Museum, Curt Teich Postcard Archives.

a new cadre of experts. And "advertisements for training equipment—blocking sleds, tackling dummies, rowing machines—were liberally sprinkled with references to science." As Hardy concludes, "the spread of standardized goods does reflect and support the spread of standardized behaviors and values."[13] It is an open question whether such developments reduced the element of true "play" in organized sports.

Along with the game itself, with its long statistical record, and the hometown players, it is the ballparks that claim the affection and loyalty of baseball fans. The earliest diamonds were unfenced, and it was only in 1862 that an outfield fence created the first enclosed park—the Union Grounds in Brooklyn, New York. From then until the end of the century, dozens of wooden ballparks were constructed and almost as many burned down—ten in 1894 alone.[14] Early in the next century, clubs turned to the steel and concrete construction technology that was transforming cities across the country. Shibe Park (later Connie Mack Stadium) in Philadelphia and Forbes Field in Pittsburgh were the first, built in 1909.

By 1923, a dozen more had been constructed. One was Redland Field, home of the Cincinnati Reds. In 1934, the owner Powel Crosley changed the name to Crosley Field to honor himself and advertise the Crosley automobile which he manufactured.

Because most of these parks were built in cities, available land was often limited. Ebbets Field, home of the Brooklyn Dodgers, was fitted into one city block. As a result, the stadium outlines were often highly irregular and seating was wrapped around the playing field like a glove. Wrigley Field, in Chicago, had been built in 1914, and between 1923 and 1926, to increase seating for fans from 14,000 to 40,000, an upper deck was added. In the '30s, capacity was increased again when outfield bleachers were constructed. The Cleveland firm Osborn Engineering was a specialist in remodeling older parks, and rebuilt Boston's Fenway Park in 1934.

After World War II, however, club owners, often with the active cooperation of city officials, tore down classic stadiums to build newer, more commodious structures surrounded with acres of parking. In 1961, there were 20 ballparks over 30 years old, and half of them were nearly 60. Ten years later, only 5 were remaining, and between 1961 and 1979, some 17 new parks were built. Some, like Atlanta Stadium, were on such a scale that no fan was closer than 100 feet to the playing field.[15] The Houston Astrodome, finished in 1964, was 18 stories high (the score board alone was 4 stories tall) and because of the local weather, was enclosed by a hemispherical roof with obligatory air conditioning. The roof originally contained Lucite windows to allow enough sunlight to reach the Bermuda grass playing field, but when players complained about the glare, the grass was replaced by bare earth painted green. The area was gradually covered with Monsanto's artificial ChemGrass, later called Astroturf because of its association with the stadium. The field was configured to accommodate both baseball and football games, as were other multipurpose stadiums built since.

Basketball is perhaps the best known American sport internationally. First played in 1891, the game was the invention of James Naismith, who was born in Canada, took degrees from McGill University and the Presbyterian seminary in Toronto, and later completed a medical degree in Denver. In 1890, he enrolled in the Springfield, Massachusetts YMCA where, the following year, the head of the gymnasium department, Luther Halsey Gulick, gave him the assignment of coming up with a game based on rules he had set down.[16] Named basketball by Naismith, it was designed to be played indoors and lent itself, with the help of electric illumination, to providing sport during cold northern winters.

It was a simple game to understand though difficult to execute. The *New York Times* commented in 1893 that the game had "within two years acquired a popularity which places it on a par with the oldest and best known of indoor sports." Not only was it played from New York to San Francisco, and in elite women's colleges such as Wellesley and Smith, but it had traveled to Japan and Australia. The newspaper helpfully explained that the game "is a modified kind of football with every element of roughness eliminated." It is played, it continued, "with a regulation association football [that is, soccer ball] and two baskets, which are fastened to the walls, if played indoors, opposite each other, and about ten feet from the floor." The baskets themselves were of "strong iron hoops with braided cord netting" although, it added, "any ordinary basket large enough for the ball to enter will answer the purpose."[17]

Naismith went to the University of Kansas in 1898, where he was made chaplain, professor of physical education, and basketball coach. He remained at Kansas until his death in 1939. In the years between, the game had spread around the world largely through the activities of the YMCA. Also during World War I the American army was staffed with hundreds of physical education teachers assigned to keep the troops busy and happy, and basketball followed wherever they went in Western Europe. Naismith reportedly was gratified by this success but remained modest about the game itself. When one of his former students announced that he had been hired by Baker University to be its basketball coach, Naismith is reported to have exclaimed, "why Basketball is just a game to play. It doesn't need a coach." In fact, by the time the original two-page, thirteen-point "Founding Rules of Basketball," written in 1891 by Naismith, was sold by his family for $4,338,500 in 2010, the original 600 words of rules had been expanded to an official 30,000.[18]

While the game's rules evolved drastically, the equipment used for play did not. By 1909, the ball and basket were largely standardized and little change followed. In 2006, the NBA (National Basketball Association) worked with the sports equipment maker Spalding to come up with a new ball covered with a synthetic composite material. The players had not been consulted and complained that the material cut their hands and that the ball did not bounce like the old leather one. In the face of nearly unanimous player rejection of the ball, it was withdrawn beginning January 1 the next year. The abortive change had been the first attempted change by the NBA in more than thirty-five years.[19]

The bicycle is a technology that refuses to be confined to any one category of play. It is a toy that is often given to children, sometimes at a quite young age, perhaps with "training wheels" attached. It can also become a hobby, with enthusiasts spending many hours taking it apart and putting it back together and many dollars on high-tech machines, colorful lycra outfits, cool sunglasses, and other accessories. It is certainly also a popular and important sport, as the Tour de France demonstrates each year. And, importantly, it has always been a powerfully liberating technology, providing people (perhaps most importantly women and children) a sense of freedom and competence, the results of which would be difficult to overestimate.

The American inventor and scientist W J McGee in 1898 claimed the bicycle was a typically American device. "Invented in France," he wrote, "it long remained a toy or a vain luxury. Redevised in this country, it inspired inventors and captivated manufacturers, and native genius made it a practical machine for the multitude; now its users number millions, and it is sold in every country." Like Frances Willard, McGee was attuned to the gender implications of the bicycle. It has, he wrote, "broken the barrier of pernicious differentiation of the sexes and rent the bonds of fashion, and is daily impressing Spartan strength and grace, and more than Spartan intelligence, on the mothers of coming generations."[20]

By 1893, when Willard had learned to ride, she and her friends were using the so-called safety bicycle. The "Penny-farthing" design, with its large front wheel and small one at the rear, was used for sport by men who gloried in its danger and difficulty. It was introduced into the United States in the 1870s, and by 1878 Albert Pope was making his own Columbia bicycle on the same model. The introduction of the pneumatic tire by John Dunlop in 1888, and the safety bicycle with its two wheels of equal size and with a chain to transfer power from the pedals to the rear wheel, opened up bicycling to the masses. Pope had adopted the arsenal system of manufacturing, with machines producing interchangeable parts for his machines. The industry grew rapidly, with people like the Duryea brothers, builders of the first American automobile, and the Wright brothers, builders of the first successful American airplane, becoming involved in the production, repair, and sale of bicycles. In 1880 enthusiasts organized the League of American Wheelmen, which reached a peak membership of 103,000 in 1898. Before the turn of the century the League had become a leading advocate of the Good Roads Movement which, while initially concerned with the safety and convenience of bicycle riders, broadened its campaign to include the needs of early motorists.

Eight years before it disbanded, the League, in a time of tightening segregation in America, voted to limit its membership to white people. In a way it was a public recognition that African Americans had taken up the bicycle, and no doubt for

Major Taylor, the African American bicycle racing champion, won many races both at home and abroad, touring countries such as Australia to earn money and publicize the sport.

the same reasons as impelled other Americans. Even the issue of female emancipation was raised by the African American author of the jeremiad *The Taint of the Bicycle* (1902). In this cautionary tale a young wife, who had joined a bicycle club, was brought to shame and eventually death by a sexual liaison with, presumably, a fellow member. For some African American men, however, the bicycle opened a rare road to fame and accomplishment. Marshall W. ("Major") Taylor, who had been born in 1878, became one of the first American international racing champions, reportedly winning 117 of the 168 races in which he competed. In 1903–1904 he toured Australia, winning races and promoting the sport of bicycling.[21]

The first bicycle craze was relatively short-lived, with usage dropping sharply during the first decade of the twentieth century. By 1920, bikes were best known as toys for children, as both they and hobbyists adopted the heavy "cruiser" styles made famous by the manufacturer Schwinn. Ignaz Schwinn, a German engineer, had begun his company in 1895 and the next year began sponsoring a racing program. His bicycles dominated the American market for most of the twentieth century, and by the early twenty-first the firm was making, according to its website, road bikes, hybrid bikes, electric bikes, mountain bikes, urban bikes, and children's bikes, as well as the classic cruisers.

The range of Schwinn's offerings reflects the continuing evolution of the bicycle in American culture. Lighter bikes than the cruisers were imported from Great Britain in the 1950s, featuring hand brakes and changeable gears. Sales of bicycles doubled in the 1960s and again in the first half of the 1970s when the ubiquitous "ten-speed" bike sparked another craze. Heavier and more rugged mountain bikes appeared in 1981 and, as the Schwinn website's home page suggests, riders tempted out of their cars in the twenty-first century have a wide range of styles from which to choose.

Some truly advanced bicycles were available to serious racing teams. In 2003, an American rider in the Tour de France rode a bicycle made almost completely from carbon fiber. This light and strong composite material was first created by American chemists in the 1960s, and found use in areas from aerospace to golf clubs and fishing rods. Indeed, the Boeing 787 Dreamliner, designed to replace the aging 747s, contains 35 short tons of carbon fiber reinforced plastic, including 23 tons of the fibers themselves. Bicycle parts, like wheels, handlebar stems, and even helmets made of the material were perhaps more common than entire bicycles, but the trend was in that direction, especially since Chinese manufacturers, which dominated the world market, were aggressively pushing forward with the fibers. Even more exotic materials, such as beryllium and magnesium, were being experimented with and innovation seemed unlikely to stop even there.[22]

Tennis was another of the "new" sports introduced into the United States after the mid-nineteenth century. The Medieval game, called "real" tennis, evolved from an earlier game played with gloves rather than racquets. The French king Charles IX granted a charter in 1571 to the Corporation of Tennis Professionals, and rules for the game were codified in 1599. Modern, or "lawn" tennis, began in England in the mid-nineteenth century and the first tennis club was founded at Leamington Spa in 1874. The Wimbledon championships were inaugurated in 1877 with men's singles. Women's singles and Gentlemen's doubles were introduced in 1884, followed in 1913 by Ladies and Mixed doubles. Tennis came to America in 1874 at the Staten Island Cricket Club, and other tennis clubs were in operation by 1881 when the United States National Lawn Tennis Association was formed to standardize rules and organize meets. The first U.S. National Championship was held that same year.[23]

While the rules and competitions for tennis were being formalized, equipment, as with other sports, continued to evolve. The engineer Frederick Winslow Taylor received a patent for a tennis racket in 1886. Existing rackets had a handle in the same plane as the head. Taylor suggested three different designs of racket with the head angled away from the line of the handle. "The object of my invention," he wrote, "is to furnish a tennis racket by means of which a certain class of players will be more readily enabled to strike balls which bounce close to the ground, so as to cause them to rise over the net, and to strike certain balls which are above the level of their heads in such a way as to drive them to the ground in the opposite court."[24] His improvement seems not to have been widely adopted, but in 1889 he tried again, receiving a patent for an improved "Lawn Tennis Net." After noting that the normal wear and tear on nets occurs mostly at the top and near the middle, he suggested three ways of strengthening the net at those points.[25]

Although Taylor's innovative designs never became popular, tennis rackets did undergo significant change as players sought a technological advantage on the court. Traditional rackets were made of wood with strings of cow intestine (gut). The use of laminated wood and better strings were about the only improvements in rackets for almost a century after 1875, and for that time they remained relatively heavy and with small heads. Then, in 1967, the Wilson Sporting Goods company introduced the first popular metal racket made famous by Jimmy Connors and other professionals. In 1975, the oversized Prince Classic was introduced, with an aluminum frame and a string area over half again as great as wooden rackets. Aluminum frames tended to warp under impact, however, and professional players needed something more rigid. The solution was a material of carbon fibers mixed with plastic resins, called "graphite." The Dunlop Max 200G, used by both John McEnroe and Steffi Graf in 1980, weighed only 12.5 ounces. In the years since, experiments have been made with other light and rigid materials such as ceramics, fiberglass, boron, titanium and others, though graphite remained the standard.[26]

Lawn tennis began with German-produced hollow balls made of vulcanized rubber, which were soon covered tightly with flannel, and it became necessary for governing bodies to standardize balls for size, weight, deformation, and bounce. White and yellow (usually bright "optic" yellow) are the only approved colors for

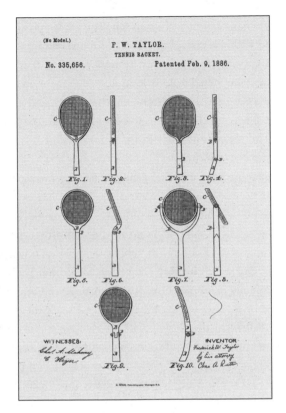

(No Model.)   F. W. TAYLOR.
TENNIS RACKET.
No. 335,656.   Patented Feb. 9, 1886.

Frederick Winslow Taylor was the putative "Father" of the efficiency craze at the turn of the twentieth century. An avid golfer and tennis player, his ingenuity sometimes pushed the bounds of the practical. Drawings for patent No. 335,656 issued to F. W. Taylor February 9, 1886 for "a new and useful Improvement in Tennis-Rackets."

professional play in the United States. The optic yellow was introduced in 1972 to make it more easily visible on color television sets. Felt has replaced the flannel of an earlier time and pressurized cans keep the more expensive balls fresh for greater bounce, a quality that quickly degrades after the balls are taken from the can. Non-pressurized (and therefore slower) balls are produced in colors such as red, orange, or green, and are recommended for children and other beginners.[27]

Like tennis, golf seems to have arisen in Medieval Europe and taken its modern form in Scotland; in 1744, the oldest surviving rules were laid down by the

Company of Gentleman Golfers in Edinburgh. The game spread throughout the Anglophone world with the rise of the British Empire, along with a little boost from a New York City newspaper advertisement for 1799 that listed golf clubs and balls, presumably imported from Britain, for sale. Again, like tennis, it began to gain wide popularity after the mid-nineteenth century, and in 1894, delegates from prominent clubs formed what was to become the United States Golf Association (USGA). There were 267 clubs in 1910, more than 1,100 in 1932, 5,908 in 1980, and in 2013, more than 10,600.[28]

The three most important material aspects of the game have been the ball, the clubs, and the course itself, and each of the three has evolved in relation to the others. The earliest balls were handmade of three pieces of leather and stuffed with feathers. These were replaced in the mid-nineteenth century with others made of gutta-percha, a kind of rubber. A third type was that patented by G. B. Work and C. Haskell in 1899. They described their ball as "comprising a core composed wholly or in part of rubber thread wound under high tension, and a gutta-percha inclosing shell for the core, of such thickness as to give it the required rigidity." Known as the Haskell ball and relatively easily mass produced, this, like the nineteenth-century gutta-percha ball before it, stimulated changes in the design and fabrication of golf clubs. The experience of players led to the observation that balls flew further once they had become nicked and dented, and this led to experiments with rough surfaces. The now familiar dimpled surface began to appear around 1910. During the 1930s, the USGA stepped in to set standards for weight and size.[29]

When the "feathery" ball was commonly used, clubs were made of local woods—in the United States, hickory for the shafts and American Persimmon for heads. The tougher gutta-percha ball allowed the use of iron for heads and, beginning in the 1890s, for shafts as well using drop forging. Both iron and wooden clubs began to be manufactured in quantity as the game of golf grew rapidly in popularity around the turn of the century.

Frederick Winslow Taylor was a golfing enthusiast of that period. When he decided to build a course at his estate near Philadelphia, he exhaustively researched and experimented with different types of grasses. Taylor confessed that "his principal reason for making these experiments was the pleasure which he derived from the investigation, still, as time went on, and useful results were attained, there came the secondary object of trying to help the golfers of the country to get better putting greens."[30] Additionally, according to his biographer, "in Taylor's case the thinking stimulated by golf, having to do with the nature and use of tools was all too closely allied with his customary thinking."

Besides research on grasses for the course itself, Taylor was interested in clubs. In 1903 he received two patents, on the same day, for the type of club called "mashies" or "niblicks." His improvements related to covering the striking surface of the head with teeth designed to give the ball a backward rotation. Similar designs were at the time being produced by other inventors. His most radical design for a club was apparently never patented, but was used by him until it was outlawed by the USGA. This was a putter, the shaft of which was attached to the head in its center and then divided into two sections about half way up toward its handles. It looked like a long capital Y and Taylor swung it forward from between his legs to strike the ball. Over a hundred years later golf's governing bodies were debating the banning of another innovative putter, the "broomhandle" or "belly putter." In use since the 1980s, this club had an extra long handle that was tucked under the chin or held against the chest or stomach and swung in a pendulum motion. In objecting to the club, the executive director of the USGA was quoted as saying that "swinging a club is the essence of the 600-year-old sport," and the chief executive of Britain's Royal and Ancient golf association charged that "anchored strokes threaten to supplant traditional putting strokes which are integral to the longstanding character of the game." The object was to preserve the "skill and challenge which is a key component of the game of golf."[31]

After the introduction of grooves to the faces of clubs early in the century, the introduction of steel shafts in the mid-1920s, the use of graphite shafts and titanium "metal woods" late in the twentieth century were the most important innovations in clubs. Computers have been used to help design both balls and clubs, since the 1980s.[32] Along with these, the courses themselves have at times had to be redesigned. The sometimes very large financial investment in courses, as well as the accumulated history, records, and tradition, were among the reasons for the reluctance of governing bodies to allow new balls and clubs, which tended to dramatically lower scores. Equipment that would allow a golfer to reach a Par 5 green in just two strokes would reduce both the challenge and the glory of competition. In 2006, the historic Augusta National Golf Club, opened for play in 1933, was criticized for a redesign that increased the playing area to 7,445 yards (six holes were lengthened) and added new trees, rough, and bunkering as well as narrowing some of the fairways. According to reports, "In an attempt to keep pace with the high-tech advances of modern golf technology, Augusta's guardians insist the facelift is part of an ongoing effort to maintain the course's integrity and ensure it will be played the way it was originally designed."[33]

Besides changes in the three major elements of the game—balls, clubs, and courses—a range of newly designed technologies became available to golfers who wanted to improve their scores without the bother of improving their games. At the low end, special gloves have been designed to improve the grip on the club, and special shoes, made of Tibetan Yak leather, provide "stability, ground feel and swing power." A divot repair tool facilitated replacing the chunk of turf torn up when hitting the ball off the tee.[34] A range of videos allows one to play a famous course alongside Tiger Woods. On some courses, golf carts are equipped with flat-screen displays utilizing global positioning systems that tell "the precise distance to every hazard, how far you need to hit past it, where the green is and precisely how far the pin is."[35]

Miniature golf links were at the other end of the technological scale of game, requiring little more than scaled-down clubs and a ball. Having evolved from genteel nineteenth-century British miniaturization of proper golf links and putting greens, the earliest facility to have all the classic ingredients of the modern miniature golf link was built in 1916 by James Barber in Pinehurst, North Carolina. Designed by an "amateur architect of fiendish ingenuity," Barber's layout was for his own pleasure and lay behind the closed gates of his estate, not available to the public.[36]

In 1922, Thomas McCulloch Fairbairn, an Englishman who wanted to build a small golf course on his Mexican cotton plantation, found grass too hard a problem to maintain in that arid environment. Instead, he and two friends from El Paso, Texas, developed a surface material made up, as his patent application described it, of "crushed cotton seed hulls with fibers thereon, and a viscous substance binding the cotton seed hulls into a homogeneous mass" which he then dyed green.[37] In 1928, Fairbairn entered into a business arrangement with Drake Delanoy and John N. Ledbetter, who two years before had begun to use his patented surface for putting greens laid out on the roofs of high-rise buildings in New York City. After agreeing to a cash payment and royalties to Fairbairn, the two New Yorkers went on to build 150 rooftop golf courses in the city.[38]

The beginning of the miniature golf craze in the 1920s, however, is credited to Frieda and Garnet Carter. It was Frieda, with some instruction in painting and a "small drafting kit," who not only designed many of the fantasy structures (including a Fairyland Gas Station) on the grounds of their Fairyland Inn on Lookout Mountain in Georgia, but in 1926 laid out a miniature golf course apparently "solely for her own amusement." It became very popular with their guests, however, and on November 6, 1930, Garnet Carter applied for patents for several "ornamental Design[s] for a Miniature-Golf-Course Unit."[39] Calling his system "Tom Thumb" golf, it was filled with hazards and statues of gnomes and elves. After having trouble keeping his grass course in repair, Carter paid Fairbairn $65,000 to use his patented cottonseed hull surface. Carter also struck a deal with W. S. and A. J. Townsend, who had become avid miniature golf players, to convert their Rochester, New York, factory from making valves for gasoline pumps to manufacturing hazards for Tom Thumb links. They employed 200 people, including a dozen artists to hand paint the fantasy features, and negotiated the exclusive rights to sell Tom Thumb links north of the Mason-Dixon Line and east of the Rocky Mountains. By 1931 they had sold equipment for 3,000 links and by the next year had paid Carter over a million dollars in royalties. The miniature golf business was relatively inexpensive to start up, and it provided relatively cheap entertainment for families hard hit by the Great Depression. In fact, Tom Thumb was deemed "Depression-proof."[40]

The 1950s experienced a revival of miniature golf, but also a schism in the ranks of the entrepreneurs who built them. Although still seen largely as a clean, wholesome entertainment for the entire family, and conveniently located in the new shopping strips or along the equally new freeways, Don Clayton wanted to return the game to a true sport, featuring athletic skill rather than good luck with the gimmicks and hazards. His course, built in 1953, featured straight putting and indeed was called Putt-Putt Golf. Along with Holiday Inn and McDonald's, he pioneered in franchising his links. "Our putters are great athletes," he wrote, "and great men. We have made competition out of a thing that was recreational. I believe this is the type of drive and commerce that made this country great."[41]

Two years later an entirely different aesthetic was adopted by Ralph J. and Alphonse Lomma, who had decided after playing a course in New Jersey that it was "too still." In 1955 they opened their own miniature golf course in Scranton, Pennsylvania. Until this time Ralph Lomma had manufactured cast-iron skillets and die-cast ornaments, but after organizing a new company, Lomma Miniature Golf, the brothers concentrated on making courses for sale, eventually prefabricating them. From the beginning, Lomma courses featured revolving wagon wheels, flashing traffic lights, paddle wheels, wishing wells, a revolving windmill, the blades of which slapped ill-timed balls away from their targets, drawbridges that opened at inopportune moments, a whale that spouted water when a ball entered its mouth, and their proudest achievement, a clown at the end of the course which ate the balls. Ralph attended to the business and dreamed up the hazards, while Alphonse "tinkered with motors, belts, gears and pulleys to create the mechanical obstacles." Over the years they sold nearly 6,000 golf courses in 337 countries.[42] As the authors of a book on miniature golf courses concluded, eighteenth-century pleasure gardens "evoked an atmosphere of exotic romance and fantasy. And so does miniature golf, as we hope it always will. For miniature golf, a truly indigenous American art form is the stuff of our collective dreams."[43]

Competitors in most sports, whether professional or amateur, are expected to wear appropriate clothing, including the right shoes. Special shoes were developed in the nineteenth century for croquet (rubber soles with canvas uppers) and the bicycle craze at the end of the century saw the introduction of heelless shoes. In 1917, Converse introduced sneakers with a higher top for playing basketball, and also in 1917, the popular sneakers called Keds were introduced and worn by children to play all manner of sports.

Running shoes of tightly fitting leather with spiked soles were developed, and by the mid-1890s the Spalding sports equipment company catalog was featuring three grades of shoe, all low-cut (with one using kangaroo hide) and using six spikes. In the 1890s also, Joseph William Foster founded a sports shoe company in England which famously provided shoes for the successful British team in the 1924 Olympics. In 1958, the company name was changed to Reebok.[44] In 1924, Adolph (Adi) Dassler and his brother, sons of a cobbler, establish the German company Gebruder Dassler OHG (later called Adidas) to manufacture his spiked soccer and running shoes. The latter gained fame in the 1928 Amsterdam Olympics and were worn by the American sprinter Jesse Owens in the 1936 Berlin Olympics. By the time he died in 1959, Dassler had more than 700 patents related to running shoes and other types of athletic equipment. His brother split from the firm in 1948 and founded Puma, which became a strong competitor of Adidas.[45] The Dunlop company, which had been established in Dublin in 1889 to make inflatable bicycle tires, established a branch in Melbourne, Australia, in 1893, and in 1924 began manufacturing rubber-soled tennis shoes there. An Australian player who worked for the firm returned from a trip to America in 1939 bringing with him a pair of yachting shoes he had bought in Newport, Rhode Island. He suggested the company reproduce the herringbone anti-slip tread on their tennis shoes and, wearing a pair, he won the Davis Cup doubles that year. Called the Dunlop Volley, the shoes proved popular until overtaken by more high-tech brands.[46]

The rapidly increasing popularity of running and jogging in the early 1970s was accompanied by an upsurge in injuries to runners; estimates were that as many as 65–80 percent of runners suffered a range of maladies from aching Achilles tendons to sore knees. In response, sports doctors joined with manufacturers to develop the modern running shoe designed to absorb the impact when the heel hit the ground. The key design innovations were to lift the heel, cushion the midsole (with air, a gel, plastic, or springs), and fortify the instep to prevent the foot rolling in.[47] In 2008 the Spira Stinger, which had a patented spring in its insole, caused controversy because the U.S. Track and Field organization prohibited "the incorporation of any technology which will give the wearer unfair advantage, such as a spring." The manufacturer challenged the rule, stating, "It's just like the battle with the oversized tennis racket or the metal driver . . . They were banned once, too. Now that's all you see. Rules can never keep pace with technological change."[48] About the same time Adidas advertised the "world's first intelligent shoe" which had "a sensor that measures the compression of the heel on impact with the ground. This data goes to a microchip in the arch that decides whether the shoe's heel is too soft or firm before instructing a mechanism that adjusts the cushion."[49] In 2006, the Nike + iPod carried a small sensor in the shoe which communicated wirelessly to an iPod Nano, informing a runner of his or her pace, distance covered, calories burned, and time spent exercising. A running shoe developed not to enhance performance but to track the wearer was introduced by Quantum Satellite Technology. It contained a GPS device to allow parents to keep track of their children, and children to track down wandering parents suffering from dementia.[50]

At least as early as 2009 research was showing, however, that shoes with reinforced arches and, especially, raised and cushioned heels not only had no effect on preventing injuries, but may actually have been making them more likely. The design, it was claimed, encouraged runners to land on the heel, which was damaging because completely unnatural.[51] When the Ethiopian champion Abebe Bikila won the 1960 Olympic marathon running barefoot, it was a powerful suggestion that perhaps expensive, highly engineered shoes were not necessary, even at the frontiers of performance. By the start of the twenty-first century it was even being argued that if running barefoot was too extreme for most recreational runners, then minimalist shoes would at least be better than regular running shoes.

Sometimes called "slippers with a grip," minimalist running shoes by 2011 were being sold (and in increasing numbers) by all but one major running shoe company: Nike made the "Free," Brooks a "Pure" range, and New Balance called its shoe simply the "Minimus." Most distinctive was Vibram, with its "Five Fingers" foot-gloves.[52] Such shoes represented the latest twist in a long record of technological innovation in sports footwear.

Clothing for competitors was also the subject of technological innovation. By 2006, traditional shorts for runners, and the elastic tights that had partially replaced them, had been joined by "compression tights." These used "strategically placed bands of sturdier fabric" designed to support the leg muscles and make them work more efficiently.[53] Some athletes used such tights in the 2008 Beijing Olympics, along with uniforms of a fabric called "Climacool" meant to cope with the expected heat at the games. According to one hurdler, "the Olympics is every athletes' dream and to be able to wear the latest technology that will help us perform at our best is very exciting."[54] An extreme example of high-tech clothing was the bionic "ReWalk" suit which contained sensors and a computer that allowed people who were paralyzed to move about in something like a normal manner. In 2012 a woman, paralyzed from the chest down, completed the London marathon, walking three kilometers a day.[55]

It was swimmers, however, who attracted the most attention with new suits. When the Australian swimsuit maker Speedo outfitted the entire national swim team for the 1956 Melbourne Olympics they created a sensation, not because of any new technology but because of the very brief cut of the men's suits. In 2003, however, the company brought out its uniquely constructed Fastskin, which utilized a Lycra-mix said to mimic characteristics of a shark's skin. The next year they brought out their Fastskin FSII, and wearing the suit, the American swimmer Michael Phelps won an unprecedented eight gold medals at that year's Olympics in Athens.[56]

Then in 2008 Speedo unveiled its new LZR Racer swimsuit woven of elastane-nylon and polyurethane. The result of truly international cooperation, the suit was developed by an Italian company (and patented in Italy) working with the Australian Institute of Sport and Speedo-sponsored swimmers. The design was developed with NASA fluid flow software, tested in NASA's wind tunnels, and it was manufactured in Portugal. The suit was approved for the 2008 Olympics in Beijing, and 98 percent of all medals won in swimming at that meet were won by swimmers wearing the LZR Racer. At the 2008 European Short Course Championships in Croatia, where some swimmers reportedly worse two suits at once to increase their effect, seventeen world records were broken. Complaints were later filed that the LZR was a form of "technological doping," and the next year FINA, the international governing body, ruled that suits must not extend beyond the shoulders and ankles and limited the thickness and buoyancy of fabrics. They wanted, they said, "to recall the main and core principles that swimming is a sport essentially based on the physical performance of the athlete."[57]

In both track and swimming events it is of obvious importance to record, as accurately as possible, who crossed the finish line, in what order, and in what times. At the other end of the contest, it is equally important to make sure that all competitors start at the same time. By 1932, stopwatches were used to time track events, with up to twenty-five persons using their watches, then averaging the times. In 1946, cameras were used to capture the finish, but the film had to be developed in a darkroom near the track, thereby delaying official results. Two years later, runners in the London Olympic Games broke a cord that activated an electronic sensor to record the winning time. In 2000, for the first time marathon runners wore microchips in their shoes, and in the 2004 Olympics in Athens, antennas placed every five miles recorded times and distances. By this time finish-line cameras were capable of taking a thousand images per second.[58]

The first semi-automatic swimming timer, the "Swim Eight-O-Matic Timer," was used at the Melbourne Olympics in 1956. Touch pads on the wall of the pool were introduced at the Pan-American Games in 1967 and first used in the Commonwealth Games in Brisbane, Australia, in 1987, recording times within hundredths of a second. Brisbane also saw the introduction of sensors in the starting blocks to record premature starts. Such was the distrust of the new technology that when three Canadians and four Australians were disqualified for too-quick starts, one timekeeper remarked that "it was so dangerous, we had to change our hotels in the middle of the night. People didn't trust the technology."[59]

The British "Hawk-Eye" system was developed in 2000–2001, using high-speed cameras and computers to track balls in play in several sports. A competing German system, called "GoalRef," uses a magnetic field and special ball for the same purpose, while six other systems were tested and found wanting by FIFA, the international governing body for soccer. GLT (goal-line technology) was approved for use in the 2013 Confederations Cup in Brazil and the 2014 World Cup matches. The use of technology to supplement the judgment of goal umpires was not universally welcomed and video replays were still not sanctioned. As one observer pointed out, however, "referees have to make split second judgments with the naked eye while millions of television viewers are treated to slow-motion replays, from different angles, which often show clearly whether the official was right or wrong."[60]

Being a British invention, perhaps inevitably Hawk-Eye was used earlier to monitor cricket matches as part of the Umpire Decision Review System. A component called "Hot Spot," invented by the Australian Warren Brennan, used infrared imaging to determine whether the ball made contact with the bat. Hawk-Eye traces the path of the ball from the bowler and can determine whether it would have hit the wicket. A third component, "Snickometer," used directional microphones to pick up the sound of ball hits, but has subsequently been abandoned. UDRS was first used in a match between New Zealand and Pakistan in 2009. In 2013, a controversy arose over an alleged incident in the Ashes match between England and Australia. In a situation strangely reminiscent of the race between offensive and defensive weapons systems, Australian batters were accused of masking the edges of their bats with silicon tape to prevent Hot Spot from working properly.[61] The technology was dropped by Channel Nine, which carried the British-Australian Tests, perhaps at least in part to save the $10,000 a day that Hot Spot charged for its four-camera system.[62] Ironically, Hawk-Eye was originally developed just to enhance the entertainment value for television audiences rather than to call official results in real time. Not surprisingly it was later used in a cricket video game to make the experience seem more "real," that is, more like TV.[63]

Hawk-Eye's use in tennis matches began with television coverage of several major tournaments before it was finally adopted at the U.S. Open Tennis Championship in 2006. Original tests the year before had been done with one high-speed camera, but in 2006, at the Hopman Cup in Perth, Australia, ten cameras were used to feed images to computers. Line-calling technologies also included systems called Cyclops and Auto-Ref, while Hawk-Eye was sometimes used as part of an IBM system called Point Tracker.[64]

Several well-known television line enhancements, like the yellow stripe showing first down in football or the K-Zone box at home plate seen on baseball, are not part of the official game, but for the 2013 America Cup yacht races in San Francisco Bay, Live Line (an "augmented reality") not only helped television and Internet viewers follow the action, but was used by officials to enforce rules. It was adopted in order to "make their sport a better show on TV and expand its popularity." For fans who wanted to see the races "in person," a large TV screen was erected on the lawn of a park overlooking the bay. As for any major sports event, the best view was often of the screen rather than the boats themselves.

Live Line showed lines on the water marking out of bounds and turn lines, as well as an on-water grid to help keep track of who was ahead and dotted lines (that seemed to melt into the water) to track the course of each boat. Stan Honey, a co-founder of SportVision Inc. which introduced the yellow football lines for

ESPN in 1998, has also been a co-founder of Etak Inc., a Silicon Valley firm that had pioneered GPS mapping for cars. Money to start the company had come from Mike Bushnell, of the video game company Atari.[65]

Just as improved technologies threatened the traditional (though always uncertain) definition of individual sports prowess, so did these new boundary and tracking technologies threaten to undermine the authority of human empires and other game officials. It had long been argued that although referees sometimes made human errors, these were as much a part of the game as the mistakes and miscalculations of the athletes themselves. The clearest winners were the television corporations who could better entertain and therefore expand the audience they sold to advertisers. For viewers, they became an expected and appreciated part of the sport itself.

There is no obvious place to draw a line between the railroads, telegraph, penny press, and growing working-class cities of early baseball and the contemporary golf carts with GPS, the high-tech running shoes, and the television-inspired use of Live Line for yacht races. Equipment has been upgraded, venues revamped, and rules of play made more complex in direct response to the industrialization and commercialization of sport. The tension between an ancient Greek ideal of individual effort and merit and the urge to gain whatever technological advantage might be had continues to help shape games and sports of all kinds. In fact, the record books and personal bests represent a tangible narrative of that tension.

Chapter Six

Extreme and (Sometimes) Impolite Sports

Some people reap more benefit from fear than others. There are different degrees of knowing what it's like.　　—PATRICK LAVIOLETTE (2011)

Perhaps the finest antidote for fear is fun.
　　—AMUSEMENT PARK SPOKESPERSON (1940S)

It has been suggested that "extreme sports" (also called action sports or adventure sports) is merely a media term for "certain activities perceived as having a high level of inherent danger or difficulty and often involving speed, height, extraordinary physical exertion, highly specialized gear and spectacular stunts."[1] In trying to articulate the "Reasons for Extreme Sport," one commentator, calling himself "maniacsportfan," wrote of people "looking for excitement who want to experience the feeling associated with living on the edge. . . . There are many reasons for extreme sports," he concludes, "but the main one is . . . to live life on your terms."[2]

To qualify as an extreme sport in popular culture, an activity should involve speed, great physical exertion, specialized equipment, and spectacular stunts. Generally, competitors perform in a less controlled environment than the highly prescribed and standardized regular sports, often at the mercy of changing and unpredictable natural environmental elements. It has been suggested that in the 1980s and '90s the term was used for sports, like skydiving and bungee jumping, that attracted adult participation, but that eventually it was dominated by activities such as skateboarding, which appealed to a younger demographic. Extreme sports often have a somewhat romanticized outlaw, or at least countercultural, aura, presumably reflecting an alienated youth that rejects the status quo and authority generally. Ironically, extreme sports as a category is at least in part the deliberate result of corporate design.

Contrary to any countercultural drive and uncontrolled natural environment, extreme sports attained a high profile, powerful definition and support, by way of the sports television channel ESPN, which organized the X Games in 1995, followed two years later by winter games. The winter games in Aspen, Colorado, in 2012, were said to have been telecast to 232 million people in 192 countries.[3] The games featured "tricks" such as front-flipping a snowmobile and double backflips on a dirt bike. Competition technologies include motocross bikes, skateboards, skis, snowboards, and snowmobiles. Some technologies were first featured then removed from competition, like inline skates, surfboards, and bungee cords.[4]

Even further evidence of the mainstream cultural nature of extreme sports was provided in 1999, when the U.S. Postal Service issued a set of four stamps showing BMX Biking, Inline Skating, Skateboarding, and Snowboarding. Their release was timed to celebrate the ESPN X Games in San Francisco. A USPS press release announced "X Games Athletes 'Stoked' About New Xtreme Sports Postage Stamps."[5]

Sports have always carried a certain degree of inherent danger, but in the late 1970s, a group of Oxford University friends formed what they called the Dangerous Sports Club. Their first publicized stunt was a bungee jump off England's Clifton Suspension Bridge. In 1979, David Kirke, a founding member of the club, took the jump dressed in striped pants, waistcoat, morning coat, and top hat and clutching a bottle of champagne. Kirke said they got the idea from natives of the Pentecost Islands. The television presenter David Attenborough had shown them, with vines tied around their legs, diving earthward in one of his documentaries from the 1950s. "It was they who gave us the inspiration for what might be called urban vine jumping," Kirke claimed. "Just adapting other people's ideas for one's own situation."[6]

Like all sports, those characterized as "extreme" have their own tools, rules, and venues. As historian David Edgerton noted in his book *The Shock of the Old*, through time technologies not only appear but disappear, sometimes reappear, and are often transformed, creating new hybrid forms, or creole technologies as he calls them. In the postwar years, some traditional children's technologies, like wagons, scooters, tricycles, and roller skates, virtually disappeared from the public consciousness. By the end of the twentieth century, however, roller skates had returned to meld with surfboards, producing skateboards (and, more recently, RipStik Caster Boards). Surfboards also morphed into windsurfers, roller skates most recently into heelys, scooters reappeared in high-tech forms, and the faithful Schwinn bicycle went high-tech as well. Even the dreary household chore of ironing had its technologies—iron and board were appropriated for the sport of Extreme Ironing.

Perhaps the oldest and arguably one of the most popular of these sports is surfing. It is widely believed that surfing on boards was developed by Polynesians, who brought it with them when they discovered and populated the Hawaiian Islands sometime about AD 450. Using indigenous Hawaiian trees, boards were cut, shaped, and finished: called *olos*, these boards were often up to 24 feet long, weighing 200 pounds. Island surfing was recorded by Captain Cook, but the following decades of European, and especially missionary, activity largely wiped out the sport, along with many other aspects of indigenous culture.[7]

At the turn of the twentieth century, however, there was a revival of the sport in Hawaii, and traveling Hawaiians carried it to various sites along the Pacific Rim and beyond. The August 18, 1888 issue of the *National Police Gazette* carried an illustration on its front cover of a woman on a surfboard, accompanied by a caption calling her "A Gay Queen of the Waves," and adding, "Asbury Park, New Jersey, Surprised by the Daring of a Sandwich Island Girl." Typically, the article inside did not further identify her, but took special note of her "dark eyes" and "bronze face," her hair which "tumbled down upon her shoulders," and the fact that her bathing dress "fitted close to the figure, the skirts reaching scarce to her knee." It did close with the observation that "she is as completely at ease in the sea as you or I on land, and the broad plank obeys her slightest touch."

Less surprisingly, taking advantage of the new American imperial vector, three nephews of Hawaii's reigning monarch were sent to St. Matthew's Military School in San Mateo, California, and in 1885, they were reported to be surfing on newly shaped redwood boards along the Santa Cruz coast.[8] It is probably not surprising that the good surf, good weather, and relative propinquity of California proved an attractive site for the early spread of the sport outside of Hawaii itself.

In 1907, George Freeth, who is credited with having, a few years before, reintroduced the style of riding boards at Waikiki while standing up, was brought to Southern California by the Redondo Beach-Los Angeles Railroad Company to create interest in water sports and hence give city dwellers a reason to take the train to the beach.[9] In 1911, Jack London published his memoir *The Cruise of the Snark*, which included a widely read description of what he called "Surfing: The Royal Sport." Stopping in Hawaii, he described seeing his first surfer standing on his board, "calm and superb, poised on the giddy summit, his feet buried in the

Surfing began in Hawaii but was carried to the United States, then other countries. It was most successfully established on the California coast in the early twentieth century, but this scene was perhaps the first instance of the sport being demonstrated in the United States. "A Gay Queen of the Waves. Asbury Park, New Jersey, Surprised by the Daring of a Sandwich Island Girl." *National Police Gazette*, August 18, 1888, cover. Courtesy of National Police Gazette Enterprises, LLC.

churning foam, the salt smoke rising to his knees, and all the rest of him in the free air and flashing sunlight, as he is flying through the air, flying forward, flying fast as the surge on which he stands. He is a Mercury—a brown Mercury. His feet are winged, and in them is the swiftness of the sea." Using a board "six feet long, two feet wide, [several inches thick, weighing, he estimated 75 pounds] and roughly oval in shape," London also learned to surf, under the tutelage of George Freeth, along with a friend he described as "a globe trotter by profession, bent ever on the pursuit of sensation. And," London continued, "he had found it in Waikiki. Heading for Australia, he had stopped off for a week to find out if there were any thrills in surf-riding, and he became wedded to it."[10]

The next year, following in the footsteps of Freeth, Duke Paoa Kahanamoku, who was en route to the Olympic Games (where he won a gold medal in swimming), stopped in Southern California and, using a redwood board, demonstrated his fabled surfing prowess and also how to shape boards. Eventually settling in California, where surf swimming was already growing in popularity, Kahanamoku inspired a shift from body surfing to board surfing and to the rise of surfing as both a practice and a culture.[11] It has been suggested that the "celebration of the beach and youth culture" was one of the major cultural shifts of twentieth-century America.[12] And, suggests another observer, "on those beaches where the best waves break, cults and minor sub-cultures have developed with speech patterns, vocabulary, conformity of dress, and social attitudes all built around the riding of waves."[13] In 1957, the book *Gidget*, which told the story of sixteen-year-old Kathy Kohner and her friends at Malibu, triggered not only a film by the same name and a television series, but a spate of other beach films, some starring Annette Funicello and Frankie Avalon and some with sound tracks by surf bands like Dick Dale and the Surfaris. This trend climaxed with the classic 1966 documentary, *The Endless Summer*. Surfing had been transformed from a local cult to a national culture.

Like so many other technologies, surfboards moved quickly from small-scale production to large, amateur to professional, casual to organized, and recreation to big business. All the while the technologies themselves continued to change, often with significant design input from the users, and often with sometimes radical transformations of styles, technique, and even venue of play.

In the years after instruction by Duke Kahanamoku, surfboard design and construction in California followed an evolution essentially independent of Hawaiian influences, and these changes drove a rapid evolution of the sport itself. For one thing, boards grew much shorter, as the "planks" of the 1930s shrank to ten feet after World War II, and in the 1960s were down to five or seven feet long.[14] Hollow boards, made from marine plywood and using waterproof glues, were introduced in 1926, and in 1932 the very light balsa wood, imported from South America, came into use. In the later 1940s, balsa wood cores were covered with laminated layers of fiberglass, while the balsa wood was later replaced with polyurethane foam.[15]

A number of other design changes and add-ons transformed the way boards were ridden. Bob Simmons is credited with the idea of turning up the nose of the board slightly—the so-called Simmons Spoon. Then a fin was added to the back, then two fins, and finally three. The first fin was introduced in 1935 and later developed by a California surfer who moved to Byron Bay, Australia; and the second and third fins were the innovations of Australian surfers, the latter in 1980. An example of Edgerton's "Shock of the Old" was reported in the *Sydney Morning Herald* earlier in 2009. Commenting that "Sydney's surf is renowned among board riders for being flat and inconsistent in summer," the article suggested that "one option for wave-starved surfers was to swap their three-fin thruster for a stand-up paddle board." Local former world surfing champion Tom Carroll said, "I learnt about them in Hawaii. It's been around in many cultures that deal with the ocean for centuries." Another local expert warned, however, that "people buying the stand-up paddle boards are not the Hawaiian water men or Tom Carrolls of the world. You're getting all these guys from [the wealthy suburb of] Mossman buying these stand-up paddleboards, going to Manly in crowded conditions and just taking out swathes of beginners every time they take off."[16]

The ankle leash, at first using surgical cord held on with a suction cup, was introduced in 1971, eliminating the long swims to retrieve boards after a wipe-out.[17] All of these changes in design, along with the shorter board itself, led to faster speeds, sharper turns, and the whole contemporary style of "performance" surfing.

"Surfboard designs were, and still are," wrote one commentator, "a combination of shaper skills and surfer inputs. . . . The shaper is really the artist and engineer who can translate thoughts into reality."[18] This rather romantic notion of the community of interest and close collaboration between producer and consumer was deeply embedded in the early years of surf culture. Shapers were often surfers themselves who made their own boards, later turning them into templates so that they could make the same board for friends, and finally for a larger market.

Board making was concentrated in Southern California where, in 1950, Hobert Alter (nicknamed Hobie and later the designer of a popular catamaran) began shaping in the family garage in Laguna Beach. His life dream, it was later reported, was "of never owning hard-soled shoes or having to work east of California's Pacific Coast Highway."[19] With his partner Grubby Clark he began experimenting with making boards out of foam and fiberglass. Their boards were markedly successful and produced in large numbers. In 2007, another shaper, Tony Martin, estimated that he had made more than 50,000 boards over the past thirty-five years. A journalist described him as a man with "X-ray vision. Only his medium is rigid foam, and inside each slab is a surfboard waiting to be liberated." Martin himself mused that "people don't understand what goes into making a surfboard. It's not something that can just be pumped out of a machine. Where's the soul in that?"[20]

The world of Hobie Alter and Tony Martin, however, was rapidly being undermined by the next iterations of technological change and industrial globalization. As late as the 1990s, an estimated 80 percent of boards sold in the United States were completely hand-shaped. By 2007, that number had dropped to 20 percent.[21] Already in the '90s Surftech, based in Santa Cruz, was having its boards manufactured in factories in Thailand. They were a sandwich of "expanded polystyrene, PVC sheet foam, fiberglass and epoxy," and, although light and durable, "their plastic feel," according to one source, "led traditionalists to deride them as soulless 'pop-outs'."[22] Not only were the boards made in Thailand and China stamped out of machines, they were computer-designed, virtually eliminating the mystique and skills of the old-line surfer/shaper.[23]

While production was being sent overseas, two technical developments in particular turned a sport peculiar to the Pacific Rim into a truly globalized phenomenon. The first was the wet suit. In 1951, Hugh Brander, a University of California physicist who had worked on the Manhattan Project, but was then consulting for the Naval Ordnance Laboratory, hit upon neoprene as an effective material from which to make wet suits for Navy SEALS. The invention was never patented because, as Brander later said, he thought "maybe fifty people in the country (would use it)."[24] In 1952, Jack O'Neill opened his Surf Shop on the beach in San Francisco. The same cold waters of Northern California that had challenged Brander also led O'Neill to make suits from neoprene, a substance which he claimed to have found carpeting the aisle of a DC-3 airliner.[25] The third putative early developers of the wet suit were the Meistrell brothers, who had moved from Missouri to Manhattan Beach in Southern California, became lifeguards, then scuba divers, and finally operators of a sports shop called Dive 'N Surf. Initially created to protect divers, they made their so-called Body Gloves—from neoprene as well, which they claimed to have discovered as insulation on the backs of refrigerators.[26] Consequently, as one enthusiast wrote, "thanks to the ongoing spirit of adventure, enhanced by the insulating pleasures of the wetsuit, surfers have tried waves along just about every surfable coastline on the planet."[27]

A parallel expansion of the sites for surfing was attributable to the wave machine. Hardly a new invention, a "wave-generator" had been installed on Hammerstein's Roof Garden in New York City during the summer of 1909. It was a part of the night club's recreation of Atlantic City in Manhattan, and in operation provided a venue for the swimming skills of Australia's international aquatic star, Annette Kellerman.[28] In recent years, however, larger and larger wave machines, of a number of different designs, have moved surfing inland across America and around the world. "'Surfing is now accepted as a requirement for any high-profile water park," one machine designer said. "Especially in the indoor parks found in colder states. They are very popular with families. Everyone wants to go surfing now, no matter where they live."[29] With 3,000 water parks worldwide, a third of them in the United States, the market for wave machines is large.

And so are some of the wave machines themselves. The Bruticus Maximus, which opened in 2005 in San Diego, cost 2 million dollars; it creates a standing wave nine feet tall, which allows riders to be completely in the curl. As with so many technologies, however, the visionary behind the Bruticus Maximus, "equal parts mad scientist, giddy surfer and shrewd entrepreneur," as one reporter called him, looked beyond the water pumps and "super critical sheet flow" to what he called the "ancillary revenue models"—a spectator area with such amenities as swinging hammocks and thatch-roofed bars, which, along with the waves, were part of his effort to "package the California dream."[30]

The world's largest wave pool is in the world's largest shopping center, West Edmonton Mall in Canada.[31] A nice postmodern example of simulacrum is provided by the 5-foot-tall artificial wave known as the FlowRider outside of San Diego's Wave House Athletic Club. "On its face," one journalist noted, "the idea of building artificial wave machines only steps away from real waves sounds preposterous. Charging $20 or more an hour to ride them recalls the old joke about selling refrigerators to Eskimos."[32] Similarly, the giant Ocean Dome complex, containing beaches, islands, rivers, and waterfalls, as well as a wave pool, was built in Miyazaki, in southern Kyushu, one of Japan's best surfing areas.[33]

One type of wave technology is embodied in what is called the Versareef. Developed by a team of New Zealand marine biologists, it is a tough rubber mat that lies on the bottom of a pool. Its configuration can be varied by computer-controlled pneumatic jacks which allow a surge of water produced by a conventional wave machine to be reconfigured into "a powerful, curling surf wave." Four different waves are generated: the Hawaiian, Indonesian, Californian, and Australian. The Californian is the easiest to ride and the Hawaiian the hardest. "Our innovation," claimed one backer, "has the potential to turn surfing into a stadium sport where spectators can watch the world's top surfers compete on an international circuit."[34]

In the years immediately after World War II, before the invention and spread of large wave machines and wet suits, surfboards came up out of the oceans, sprouted wheels, and evolved into skateboards. The commonly accepted prehistory of skateboards is represented by the fruit box scooter, built by children in the first half of the century. A roller skate was pulled in half, then each half nailed to either end of a two-by-four, on the top of which was fastened an up-ended fruit box with a wooden handlebar running across its top.[35]

The story continues that in Southern California after World War II, surfers, in bad weather or after the waves had dropped, used these, without the box and handlebar, to street surf. Industrially manufactured boards were sold in stores by 1959, and in 1963 professional-grade boards were available.[36] The most celebrated group of skateboarders, who took the practice from a recreation to an extreme sport, was the notorious Z Boys of Dogtown, which was a largely run-down, working-class area of Santa Monica running south through the suburbs of Ocean Park and Venice. One journalist captured the culture nicely: "Outside the Jeff Ho & Zephyr Productions Surf Shop, in front of a wallsize mural of co-owner Ho surfing a wave that was almost pornographic in its perfect, arcing glassiness, Alva and Adams and a few other Dogtown kids were skateboarding back and forth, cutting off cars, catcalling passing girls, staring down all the pedestrians who failed to avoid them."[37]

At some point in the early 1970s, the Z Boys discovered schoolyards. Cut out of hillsides, with embankments and retaining walls, "the schoolyards' asphalt 'waves' broke beautifully every single day, all year round, creating entirely new possibilities for the sport," as one of their chroniclers put it. "The Dogtown kids started applying their surfing techniques to concrete."[38] Of equal importance was the discovery of Southern California backyard swimming pools, emptied by the drought in the summer of 1976. The rounded bottoms, passing from shallow to deeper, provided the perfect concrete forms within which to create a complex aesthetic of movement called vert skating.

A collapse of the California housing market in the first decade of the twenty-first century created new opportunities. It is reported that skateboarders are using Google Earth satellite images to locate abandoned swimming pools in the state's San Joaquin Valley, the site of "rampant home repossessions." One twenty-seven-year-old located the pools, then drained them with a gas-powered pump. He claimed that "skaters had traveled from as far away as Germany and Australia to practice their aerial stunts in emptied suburban pools." Since local health officials worry that abandoned pools may breed mosquitoes and become sources of West Nile virus, the enterprising skateboarder claimed he was "doing the city a favor" by draining them.[39]

In the mid-1970s, kids without access to empty pools (or the skate parks that were beginning to spring up) built home-made "vert ramps." They were described as being "a plywood-coated two-by-four-framed structure that mimicked the transition and lip of backyard pools, but lacked the bowled horizontal curve that allowed for carving." This was considered not to be a disadvantage, since the new maneuvers, or tricks, were vertical. By the early 1980s, it was claimed, "vert skating was predominantly done on ramps."[40]

A somewhat similar advance to vert skating occurred in 1978, when Allan "Ollie" Gelfand discovered that by stamping down on the back of his board he could go airborne. Called the "Ollie" in his honor, the maneuver enabled street skating, "like parkour . . . , [to make] use of the urban landscape in creative ways. Tricks are performed on benches, hand rails, retaining walls, picnic tables, over sets of stairs, shopping carts and parked cars. And that's just getting started. For the street skater, virtually anything is rideable."[41]

As one of the Z-Boys explained, "two hundred years of American technology has unwittingly created a massive cement playground of unlimited potential, but it was the mind of 11 year olds that could see that potential."[42] So dramatic was this appropriation that the architectural historian Iain Borden was led to comment that "the emergence of streetskating in the 1980s and 1990s is seen to derive from the possibilities of modern architecture, leading to new ways of editing, mapping and recomposing the city."[43]

This particular evolution of skateboard riding was made possible by a rapid series of technological changes in all three parts of the board itself: the deck, the trucks, and the wheels. The deck came to be made of veneered wood, laminated and glued, placed in a hydraulic press to turn up the front and back, then shaped. Over time the decks were widened for better control. The steel roller skate wheels were replaced, first with clay and then in 1973 with polyurethane. More than 40 million skateboards were sold in one two-year period in the 1970s.[44]

Like wave machines in water parks, skateboard parks, with their bowls and ramps simulating the experience of the empty swimming pool, which itself simulated the waves of the ocean, sprang up around the country. These parks began to appear in the late 1970s, with one of the first being built in Florida in 1976.[45] Mostly privately built, many were largely abandoned in the '80s because of a combination of escalating liability insurance and soaring land values. In the 1990s, California passed legislation (renewed in 2003) that eased the liability burden on parks. By 2007, there were nearly one hundred in Northern California alone.[46]

The building of skateboard parks was not immune to the romantic user/builder aspect of surfboard fabrication. In the summer of 1990, a group of skaters in the Portland, Oregon area, all in their late teens and early twenties, began to construct an illegal park on 9,000 square feet of former asphalt parking lot, under a four-lane highway overpass. Because the skaters drove off the homeless, drug addicts, and prostitutes who frequented the place, local authorities ignored the park being illegally constructed. As one observer wrote, "these guys were blue-collar kids and builders, so it was a natural extension for these hard workers to become skatepark designers and engineers. After all, they had been building ramps since they were kids."[47] The park was completed in 1994, and its builders went on to form the Dreamland Skateparks company. Their website home page lists fifty-eight finished parks already built, including locations in Italy, Austria, and the UK, as well as twenty-five others still in the design stage. The company began with eight skaters and their families: "Dreamland employs no common 'laborers'" they bragged. "Every team member is a highly proficient skateboarder, and the results stemming from a passion for the sport and uncompromising focus on quality are evident. . . . Each Dreamland profile [of a team member] reveals a lifelong passion and dedication to skateboarding. Indeed, skateboarding is our life."[48]

But here, too, scaling up has taken place over time. The most extreme ramp in 2006, navigable by only two dozen skateboarders in the world, was the Mega Ramp just north of San Diego, 360 feet long and 75 feet high at its apex. One hundred eighty feet down the run was a 70-foot gap over which the rider was airborne.[49] Inevitably, one further venue for skateboarding is the virtual. The video game *Skate*, released in 2007, was said to be "far more realistic than what had come before. It was also extremely challenging: even performing basic tricks," according to a reviewer, "felt satisfying and rewarding." A sequel, *Skate 2*, added new layers of difficulty. As the reviewer notes, "in a game this difficult, there are many occasions when your character suffers a painful accident. The most spectacular tumbles are celebrated in a new Hall of Meat mode, which encourages masochists to seek out towering locations to hurl their skater into the abyss."[50]

Much of the evolution of surf- and skateboards is a tale familiar to the history of technology: the creation, then appropriation, of indigenous cultures along the vectors of imperial commerce, of industrialization overwhelming craft skills in manufacture, the globalization of popular culture, the embedding of particular tools and machines in a web of meaning. Of course, these are also all familiar manifestations of modernity. At the same time, the simulacrum of wave machines at the beach and the parkour-like revisioning of the urban landscape by the Z-Boys suggest more than a hint of postmodernity as well.

The ironic (a signal mode of the postmodern) sport of Extreme Ironing began modestly enough in 1997 when Phil Shaw of Leicester, England, torn between his desire to go rock climbing and the need to iron his laundry, compromised by moving his board and iron into the backyard. Encouraged by friends, he expanded this activity to ever more extreme sites, combining, as he said, "the thrill of extreme sport with the satisfaction of a well-pressed shirt."[51] Records were set, then broken, for ironing on the highest spot (atop a craggy mountain) and lowest (deep in the ocean). The link between extreme ironing and surfing is not entirely fanciful. As early as 1912, we are told, a surfboard brought from Hawaii to Australia, which was deemed "impossible to ride," was later used as an ironing board.[52] Unlike skateboarding, extreme ironing bends rather than reinforces standard gender categories (although it must be said that one of the Z-Boys was a girl).

The evolution of boards, out of the water and up on to land, is an almost Darwinian example of evolution, recapitulating that of life itself. Not surprisingly, boards left in the water and those that pulled up onto dry land both continued to evolve into new forms, making possible the formation of new sports. One persistent idea was to attach a sail to any sort of small craft, such as surfboard, skateboard, sled, or iceboat. Windsurfing was a relatively early example of this, though the story of its origins and development are typically complicated.

In 1965, S. Newman Darby, an artist and sign painter in Pennsylvania, published an article in *Popular Science* describing a sailboard of his own design. It had the key elements of a windsurfer: "a hand-held sail rig fastened with a universal joint to a floating platform for recreational use," and may have been the outgrowth of an idea he had been thinking about since the late 1940s. He and his brother began manufacturing the boards, but never applied for a patent.[53]

Meanwhile, in Southern California in 1962, Jim Drake, an aeronautical engineer working on such projects as the X-15 rocket plane and the Tomahawk Cruise Missile, began talking to various friends about the possibility of building something like what came to be known as a windsurfer. About 1967, Drake, an avid sailor, with a close friend Hoyle Schweitzer, an avid surfer, decided to actually attempt to build a prototype. Schweitzer, with more enthusiasm than technical skills, pushed the project forward while Drake, with an engineer's knowledge and careful consideration, proceeded somewhat reluctantly because, as he later wrote, "I didn't have the whole concept in mind particularly." Once the board was built, Drake turned to the still-unresolved problem of control. Abandoning the use of a rudder and centerboard, he devised a moveable sail anchored to the board by a universal joint.[54] The craft also had an asymmetrical sail controlled by a wishbone boom. The key element of a universal joint had been present in Darby's board, but Drake insisted that at that point he had not yet seen the *Popular Science* article. He later conceded that he was probably the third, not the first, inventor of the craft but insisted that, as an engineer, "my contribution was to make the thing actually efficient and workable."

Drake and Schweitzer filed for a patent on March 27, 1988 and received patent number 3,487,800 on January 6, 1970. The board was called simply a "Wind-Propelled Apparatus," in which "a mast is universally mounted on a craft and supports a boom and sail." They claimed that "the present invention provides wind propulsion means for a vehicle that adds new dimensions of wind responsiveness and speed and yet enhances the vehicle's normal ride and control characteristics to greatly increase the enjoyment obtained therefrom."

In 1970 the two set up a company, Windsurfing International, to manufacture and sell the boards. Drake continued with his career as an aerospace engineer, while Schweitzer and his wife worked full time on the company. In 1973, he bought out Drake's share and continued an aggressive effort to dominate the market for such boards. Rival companies, particularly those in Europe, used Darby's priority to dispute their patent, and ensuing court battles were long and numerous. All the while windsurfing as a sport developed rapidly, with a professional World Cup tour being formed in the 1980s and the sport receiving Olympic status in time for the 1984 games in Los Angeles. At the same time windsurfing designs were evolving (Drake continued to innovate in his spare time) and speeds were increasing; a record of 46.4 knots was set in 2003. As with surfboards, manufacturing largely moved to Southeast Asia, principally Thailand.[55]

In 2008, a former Olympic champion windsurfer declared that he had abandoned that sport for a new one—kiteboarding. "It's something new," he was quoted as saying, "like a new toy or a new girlfriend. It's cool to be at the beginning of a sport. It hasn't gotten too hard-core yet."[56] The idea was to "ride a small surfboard while harnessed to an inflatable nylon crescent-shaped kite." The sport was so new that there were "no restrictions on boards, kites, materials or designs: In kiteboarding, we're still just getting things figured out," one promoter noted. At a race on San Francisco Bay in 2008, it was observed that, for example, on the boards "some of the fins . . . were made of fiberglass. Others of resin or carbon fiber. Some were a foot long and skinny, some were 4 inches long and squat. Some boards had four fins, some only two."[57] It seemed to be a clear case of cut-and-try development by users who designed and built their own equipment.

Boards evolved out of the oceans onto land, not only in the form of skateboards but also, most notably, into snowboards as well. Though it is likely that individuals had been experimenting for years with riding down snowy slopes on a single board, what came to be called snowboards were first developed in the United States during the 1960s. Some of the innovators had technical backgrounds, most seem to have been surfing enthusiasts, and a few went so far as to try their hand at manufacturing boards of their own design.

One early innovator was Sherman Poppen, an engineer from Muskegon, Michigan. His inspiration came on Christmas Day, 1965, while "tinkering in his garage." His idea was to amuse his daughters with a new device made from a pair of children's skis, bound together and with a rope, or "balancing lanyard," attached to the tips. There were no foot straps, but he had attached "anti-skid footrests," along with an upturned front and back, a plastic cover on the bottom and metal protection on the sides. Poppen called it a "surf-type snow ski," a description from which his wife coined the term "Snurf."[58]

In his patent description, Poppen claimed that his aim was to create a new winter sport based on three summertime activities: surfing, skateboarding, and water skiing with the single "slalom" board. "The prodigious entertainment and athletic recreation provided by the sport of surfing is well known," he wrote. In addition, "another well known sport, recently popular among physically agile youngsters, is that of the skateboard." Indeed, he claimed to have added "the slight curl at the back of the new ski, upon which the skier may push to pivot the front portion of the ski off the ground (as is done in skateboarding) to execute sudden, exhilarating sidewise maneuvers." The Snurf proved extremely popular. Poppen signed over the manufacturing rights to the Brunswick Corp., and more than 750,000 boards were sold over the next 15 years.[59]

During the 1970s various changes were made to the basic Snurf design, leading to the modern snowboard. Dimitrije Milovich, who is credited with starting the first "modern" snowboard company, took out a patent in 1974 for a "Snow Surfboard with Stepped Stabilizing Sides" which featured "a plurality of inverted longitudinal stabilizing steps" along with "a longitudinal stabilizing skeg on the bottom surface of the body and a pair of turning skegs extending from the bottom surface of the tail section." It was Milovich's belief that previous boards had not really allowed a rider to reproduce true surfboarding moves. His own board, he claimed, "may be controlled by the user with weight shifts using techniques similar to those employed in controlling water surfboards." It was also possible, he claimed, to ride his board sitting, kneeling, or lying down as well as standing upright. Milovich wrapped up his by then struggling company in 1982, but by 2010 was said to be running a successful engineering business.[60]

Jake Burton Carpenter (or Just Jake Burton) was one of a number of competitors who became inventors and then manufacturers. He began snurfing at the age of 14, and in 1977 entered a modified board of his design in a competition in Michigan. His major innovation seems to have been attaching binding to hold feet on the board; though at first controversial, he went on to manufacture and sell large numbers of his Burton Snowboards. Changes in design and materials came rapidly during the 1970s and '80s. Various styles of use tended to lead to specialized techniques and equipment.[61]

Like so many new sports, snowboarding quickly developed organized competitions, with the first National Snow Surfing Championships being held in 1982. A new journal, the first issue of which was called *Absolutely Radical*, appeared in 1985 but immediately changed its name to *International Snowboard Magazine*. Through the late-'80s ski resorts, after first banning snowboards, gradually began to accommodate them until, by 2010, only three North American resorts still refused to allow them on the slopes. The ultimate recognition of success came in 1998 when snowboarding was added to the winter Olympics in Nagano, Japan.[62]

Boarding on snow had become one of the most popular winter sports, but boarding on rock was more of a niche activity. Lava-sledding was an ancient Hawaiian tradition—the Bishop Museum in Honolulu has an 800-year-old board on display, and one enthusiast, Tom "Pohaku" Stone, a retired lifeguard and champion surfer, has counted fifty-seven historic rock slides still visible around the islands. It was the similarity to surfing that attracted him to the practice, but the "extreme sport" experience clearly was also important. "You can't even imagine what it's like to be headfirst, 4 inches off the ground, doing 30, 40, 50 miles an hour on rock. . . . It looks like you are riding just fluid lava. It's death-defying . . . but it's a lot of fun." Before missionaries banned the sport in 1825, Hawaiians used boards often 12 feet long and 6 inches wide, carved from either kauila or ohia trees. The track was hard lava, but sprinkled with leaves for the descent, which was made standing, kneeling, or laying down on the sleds.[63] The parallel to surfing was obvious.

In fact, whether surfing was an actual inspiration or more a source of borrowed legitimacy, the parallel was drawn beyond skateboards, windsurfers, snowboards, and lava boards. In 1973, the publication of *Hang Gliding—The Basic Handbook of Skysurfing*, written by the recreational parachute diver Dan Poynter, made the connection explicit. Hang gliding had a long history before its rapid rise in popularity in the 1970s. Gliders had been a common research tool of early aviation investigators trying to discover the aerodynamic characteristics of human flight. The German engineer Otto Lilienthal made 5,000 flights in his gliders of various design between 1891 and 1896, sometimes staying in the air for five hours. His are sometimes called "the world's first hang gliders."[64]

Carl S. Bates, an Iowa aviator, built his first glider as a teenager in 1898. In 1909 he published a description of, and plans for, a bi-winged glider in *Popular Mechanics*. Four years later, it was reprinted in the book *The Boy Mechanic: 700 Things for Boys to Do*. Bates described his device, made of cloth stretched over 41 ribs plus rudders, as "a motorless aeroplane," and flying it as "the most interesting and exciting sport imaginable." To "make a glide," he wrote, "take the glider to the top of a hill, get on between the arm sticks and lift the machine . . . , run a few steps against the wind and leap from the ground. You will find that the machine has a surprising amount of lift . . . and you will go shooting down the hillside in free flight."[65]

The career of Volmer Jensen, who built two hang gliders around 1930 using Bates's *Popular Mechanics* plans, illustrates the porous boundaries between gliders (sail planes), hang gliders, and ultralight aircraft. In 1937 he moved to Glendale, California, and began designing a two-engine executive aircraft with a partner. During World War II, when private aircraft were banned in Southern California, Jensen "thought it would be fun" to build another hang glider, this time one with three-dimensional controls—the first, he claimed, to be so equipped. "I installed elevators and ailerons," he wrote, "controlled by the right hand with a device shaped like a + sign and a rudder by the left hand, using a short rubber bar. It flew great." It was an effective technological substitute for the normal shifting of body weight to control flight. "As soon as one's feet leave the ground by 12 inches," he enthused, "one feels like he's 1,000 feet in the air. This is a very exhilarating experience and a safe sport." In 1971, Jensen offered blueprints and photographs of his hang glider for fifty dollars. Materials would cost about $400, he reckoned, but he made it clear that he would not supply either materials or a kit. In the mid-1950s, Jensen designed and built his VJ-22 two-seater amphibian plane and later fitted out his VJ-23 hang glider, first with an 8-horsepower snowmobile engine and then with one of 15 horsepower. The VJ-23 was called the "progenitor" of modern ultralight aircraft.[66]

Gertrude and Francis Rogallo were also prominent innovators, applying for a patent on a "Flexible Kite" in 1948. "It was an object of our invention," they wrote, "to provide a kite of simple and economic construction and wherein the use of reinforcing members may be ordinarily eliminated." The patent was granted in 1951 and the next year they applied for another to cover improvements. Granted in 1956, this second patent was said to describe the construction of a "completely flexible kite of a single piece of suitable material which will be exceedingly cheap to manufacture and prepare for flight and which will be easily controlled in flight as well as effecting good flying characteristics." They asserted that "the basic novelty and construction of our kite is not limited to uses as a toy, but could, in fact, be expanded to more serious purposes such as radar targets, parachutes, underwater tow devices, aircraft tow targets, and so forth."[67]

Francis Rogallo, who had earned a degree in mechanical and aeronautical engineering at Stanford University in 1935, was working at the National Advisory

Committee for Aeronautics (NACA), but that agency appeared to have no interest in his kite. With the coming of the space race, however, and the absorption of NACA into the new National Space and Aeronautics Administration (NASA), Rogallo's kite was seen as a possible way to facilitate the reentry of space capsules into the Earth's atmosphere. It was eventually decided, however, to use the more familiar parachute. Rogallo retired in 1970, moved to Kitty Hawk, North Carolina because of its association with the first flight of the Wright Brothers, and took up hang gliding at the age of 62. He made his last launch off the dune on his eightieth birthday.[68]

The first person to build a glider using the Rogallo design was said to be Barry Palmer of Sacramento, California, who had read about the design in an issue of *Aviation Week* in 1961 and made his first flight that same year. Rogallo's paraglider, as it was sometimes called, also spurred interest in Australia. John Dickenson, an electronics technician, saw a picture of Rogallo's invention in a magazine and in 1963 used it as the basis for a kite for water skiers, to be pulled behind a power boat. Two of his friends, also avid water skiers, embraced the new device, and one of them, Bill Bennett took it to the United States. Bennett established the DELTA Wing Kites and Gliders company in California in 1969 and began their manufacture.[69]

In 1973 and 1974, Bennett, and his chief designer Richard Boone, began experimenting with redesigns of the Rogallo glider, developing what they called the Phoenix series. According to the Smithsonian Institution, Bennett's gliders should be considered the second generation of such devices and were important in the spread of the sport during the 1970s. A consummate showman, Bennett brought attention to hang gliding with his many stunts, such as setting altitude records, launching from a hot air balloon, and acting as a stunt pilot double for Roger Moore in the James Bond film, *Live and Let Die*. He was killed in 2004 while piloting a powered hang glider in Arizona.[70]

While boards and kites were spinning off new uses and associated sports, the practice of parkour was emerging as an activity which, as one observer put it, was "a new way of bringing together the spring-loaded bodies of young males with stairways, park benches, the stud of city blocks and suburban shopping malls." It was, he wrote, like "skateboarding, but without the skateboards."[71] Based on an amalgam of military obstacle training, martial arts, gymnastics, and similar energetic exercises, the basic goal of *traceurs* (those who participated in the sport) was to proceed from one point to another, usually in an urban setting, in as efficient and direct a line as possible; climbing walls, hurdling stairwells, vaulting rails, jumping from rooftop to rooftop. In the spectacular documentary *Jump London* (2003), Free Running (as parkour is also called) featured traceurs "turning London landmarks into their own jungle gym."[72]

As the words *parkour* and *traceurs* suggest, the origins of the practice were French, emerging probably in the 1920s as a form of physical training for the French military borrowed, at least in part, from observations of the athleticism of indigenous Africans. In the late 1980s, a handful of urban French youth, partly inspired by Bruce Lee's martial arts films, began to shape the practice of parkour, which traveled to London and then to the United States. Despite its imaginative use of the built urban environment as its "equipment," some practitioners resisted the comparison with skateboards, not wanting to be identified with the negative connotations of "rebellious and misguided youth."[73] At the same time, although admitting parkour "does carry inherent risk," a leading traceur, Dan Edwardes, maintained that it "is not an extreme sport," and as of that time had not yet been subjected to organized competition.[74] "Competition," wrote another, "pushes people to fight against others for the satisfaction of a crowd and/or the benefits of a few business people by changing its mindset. Parkour is unique and cannot be a competitive sport unless it ignores its altruistic core of self-development."[75]

Many of the categories of technology and play, like toys and hobbies, have imprecise boundaries which make precise definition impossible. Extreme sports, however, seem to suffer more than usual from this problem. An ultra-marathon over extreme terrain may clearly qualify, but at what point between that and the morning jog does the term extreme become appropriate? Bicycling is a hobby for some, a serious organized and professional sport for others, but an extreme sport when it involves BMX biking. And, to take up the challenge of the Dangerous Sports Club, is fear a necessary component of an extreme sport? One definition, variously attributed to Ernest Hemingway and others, is that really "there are only three sports: bullfighting, motor racing, and mountaineering; all the others are merely games."[76] Perhaps it is that note of masculine bravado that comes closest defining the category of extreme sport.

The historian Wolfgang Schivelbusch gave his book *Railway Journey* the subtitle *The Industrialization of Time and Space in the 19th Century*. The railroad, he wrote, was the "essential agent of the transformation of landscape into geographical space." It "was experienced as a projectile, and travelling on it, as being shot through the landscape—thus losing control of one's senses."[77] The landscape scholar J. B. Jackson noted something similar with regard to extreme sport (though the term had not yet been invented). Activities like water skiing and white-water rafting, he observed, created a "sense of danger or at least of uncertainty, producing a heightened alertness to surrounding conditions," but without allowing "much leisure for observing the more familiar features of the surroundings." The landscape in fact becomes an almost abstract background, "seen at a rapid, sometimes even a terrifying pace."[78] Extreme sports were absolutely dependent on the landscapes that gave them play. But those landscapes were transformed by abstraction, just as had been the case when the railroad came to represent the very epitome of modern travel. Sensation was increased and contemplation reduced.

Chapter Seven

Electronic Games

Spacewar could be found on nearly every research computer in
the country. —MARY BELLIS

T he story of computer games is full of ambiguities, often posed in terms of
binary choices. Is it an art or a technology; should a program be patented or
copyrighted; does it stimulate creativity or passivity in players; do games stimulate or discharge aggression and violent feelings; are games hopelessly masculin-
ized or gender-neutral? We can gain some help in thinking about these questions
by looking at the history of gaming, the individuals involved, the patent battles
that embroiled them, and the cultures that developed around the subject of gaming. And, most important, although video games appear to be at the very cutting
edge of high-tech play, they have a long prehistory of antecedents.

The precursors of video games can be traced back to the Chateau de Bagatelle
where in 1777 a party was given for the French king, Louis XVI. A highlight of
the festivities was a new table game that came to be called bagatelle after the
Chateau, the word appropriately meaning a trifle or a decorative thing. The game
was played on a modified billiard table, tilted up and away from the player, who
struck an ivory ball with a cue, avoiding a set of upright wooden pegs placed at the
far end. The game found its way to America and gained enough popularity that
an 1864 Currier and Ives cartoon featured President Abraham Lincoln playing
against his Democratic challenger, George B. McClellan.[1]

In 1871, a small but important improvement was made in the playing table
when Montague Redgrave, an Englishman who had settled in Cincinnati, Ohio,
took out a patent for a table that added "a tension spring to the piston that propels

the ball, whereby any desired quantity of force may be given and easily graduated by the eye," thus dispensing with the separate cue. "Muscular power," he claimed, sent the ball up the inclined table, and gravity brought it back down. As the ball rolled back down the table it hit obstacles, some of which had bells to mark the impact. In 1898, Redgrave patented an improvement that enclosed the piston within the walls of the table. This was because, as he said, the old protruding knob was "in some danger of being caught by passing garments, may easily be broken by children through carelessness or in their play, and will impede packing for storage or transportation."[2] Redgrave's bagatelle table is considered the first modern pinball machine.

Coin-operated versions of bagatelle, built as tables with a glass surface, were being produced by the early 1930s when David Gottlieb's 1931 Baffle Ball became the first best seller with more than 50,000 units sold to drug stores and taverns where Depression-era patrons could hit five to seven balls for a penny. One of his distributors, Ray Moloney, set up his own company and sold the popular "Bal-lyhoo" game. Eventually Moloney changed the name of his company to Bally to capitalize on its popularity. By the end of 1932, some 150 companies, mostly working out of Chicago, were making pinball machines, though by 1934 factors such as the Depression and intense competition had reduced that number to only fourteen.[3]

Technical changes made during the 1930s resulted in creation of the pinball machine celebrated in popular culture and also paved the way for the next generation's video games. Most of the changes took advantage of electricity to provide lights and electromagnetic devices to eject balls from bonus holes. In 1947, the game Humpty Dumpty introduced player-controlled flippers, which built the element of skill into the game.

The electromechanical controls of the 1950s and '60s were replaced with circuit boards and digital displays in the '70s. Hot Tip, introduced in 1977, was the first solid-state electronic game. By this time, however, video games were already invading bars and arcades, signaling the end of the pinball era. Some manufacturers of pinball machines, like Bally, continued to produce them, introducing new solid-state technology to appeal to players, but also diversified into making video games themselves. In an example of art imitating art, the game Video Pinball was written for the Atari 2600, and in 1983, Pinball Construction Set, written for the Apple II, allowed players to first build their own pinball machine, then play with it.[4]

The early years of video games as such is tangled and contested, with a number of persons being put forward as the real "inventor" of the genre. As early as 1947, Thomas T. Goldsmith, Jr., and Estle Ray Mann patented a game to be played on a cathode ray tube. Eight vacuum tubes produced a simulated missile that could be aimed at a target by manipulating several knobs. Five years later, A. S. Douglas earned a doctorate at Cambridge University on the subject of human-computer interactions. His dissertation was illustrated by a tic-tac-toe game to be played against the machine, an EDSAC vacuum tube computer, also with a cathode ray display.

Most investigators agree that in 1958 William Higginbotham, working at the Brookhaven National Laboratory (BNL) in Upton, New York, in an effort to give visitors an interesting experience to liven up the static displays on view, developed a tennis game to be played on an oscilloscope attached to a small analog computer. Players "hit" the ball (a glowing white spot) by pressing a button, and directed the return by turning a knob. By all accounts the game was very popular for the two years it was available: according to one scientist, "people would line up for hours to play it." Higginbotham neglected to patent the game, claiming, "I considered the whole idea so obvious that it never occurred to me to think about a patent."[5]

There is also wide agreement that a more lasting impact was made by Steve Russell in 1962 at a laboratory at the Massachusetts Institute of Technology (MIT) in Cambridge, Massachusetts. The game, called Spacewar, was the work of a team of six programmers, led by Russell, and was finished in the same month the American astronaut John Glenn made his first orbital flight. Making use of the laboratory's new Digital Equipment Corporation (DEC) PDP-1, a mainframe computer with a cathode ray display, two spaceships, fighting each other against a background of stars, required quick reactions, much like what came to be called "twitch" games.[6]

Unlike the other games mentioned, Spacewar had an afterlife. The MIT group, which shared an interest in science fiction literature, continued to modify the game in the following months, changing it from a simple object-in-motion program to a complex contest involving gravity fields, stores of torpedoes and fuel, and ships that could disappear and reappear. Again, the game was not patented, and one of the team later claimed that "the only money I made from Spacewar was as a consultant for lawsuits in the video game industry in the 1970's." In keeping with the developing hacker culture of the times, however, paper tape copies of the program were made and distributed for free, finding their way to other university laboratories around the country. Meanwhile, of the six programmers responsible for the game, one joined the National Security Agency, two went to work for DEC, one moved into artificial intelligence, one into researching neural nets for General Motors, and one went to work for a software company.[7]

Beyond the search for starting dates and "first" inventors, confused partly by the contested definition of "video" or "computer" or "digital" games, looms the cultural context for these "firsts," a culture that linked "university computing departments, the military, the interests of the first game developers, the first games and the subsequent development of game playing as an activity embraced largely by young males." It was claimed that in the mid-'60s, *Spacewar* could be found on nearly every research computer in the country." In a world of "nuclear angst and consumer confidence," of drive-in cinemas, science fiction, and rocket fins, it was a "highly masculine world . . . that . . . shaped the subsequent development of game genres, public game spaces and digital games as a cultural activity."[8]

At the same time, the first games were produced in sites familiar to historians of both art and technology. As historian Thomas P. Hughes has pointed out, "inventors were like avant-garde artists resorting to the atelier or the alternative life-style. . . . Working in their retreats, intellectual and physical, they created a new way, even a new world." "The inventors," he continues, "created machines and processes among which they felt at home, and the artists invented pure and ordered spaces filled with music, painting, and sculpture."[9]

The commercialization of video games began almost as a cottage industry. As Stewart Brand, later founder of *The Whole Earth Catalog*, noticed when he came upon a group of students playing Spacewar in a laboratory at Stanford University, "they were absolutely out of their bodies, like they were in another world. . . . Once you experienced this, nothing else would do. This was beyond psychedelics."[10] One Stanford graduate student named Bill Pitts created a coin-operated version called Galaxy Game which was operated for more than six years. More significantly Nolan Bushnell, a graduate student at the University of Utah where he saw Spacewar being played, in 1971 created an arcade version he called Computer Space. The game was created while he worked for Nutting Associates, a manufacturer of pinball machines. The analogy between their usual product and an arcade video game was obvious, but because the latter was much more complicated to play, it was not a commercial success. Bushnell quit the firm, founded the iconic Atari company a year later, and introduced the classic arcade game Pong.[11]

Another early pioneer was Ralph Baer who in 1951 had gotten the idea for a video game to be played on a home television set. His employer at the time was not interested in the idea, but he returned to it in 1966 and produced several games characterized as chase games, ball and paddle games, as well as shooting games. By this time he was working for the Equipment Design Division of Sanders Associates, a military electronics firm. He worked on his games without company authorization, then, when he had a prototype, demonstrated it to the corporate director of patents and the director of research and development. As Baer later wrote, "playing chase games and shooting at the screen turned them on." Development continued, patents were taken out, and in 1971 he drew the interest of the television builder Magnavox, which placed their first game, Odyssey, on the market in 1972. According to Baer, he thought of himself as "just a guy with a TV engineering degree who wanted to do something with all those 60 million TV sets out these besides tuning in Channel 2, 4, or 7."[12]

The importance of Baer's patents was demonstrated when Bushnell's Pong became a huge success. The fist machine, handmade in his apartment, was set up in a tavern in California where it was soon out-earning the nearby pinball machines by a factor of six to one. Bushnell's new Atari company released its first commercial Pong in late 1972 and eventually sold ten thousand of them. Four years later he sold Atari to the motion picture company Warner. As it turned out, the firm was sued by Magnavox for infringing Baer's patents, and as a result of its success, Atari had to pay a license fee on every set it sold.[13]

It was estimated in 1983 that during the previous year, $200 million worth of software packages for home computers were sold in the United States, of which some $106 million went for "recreational products," that is, games. That same year customers (largely young people) spent a staggering $1.4 billion in the United States playing video games in arcades, pizza parlors, and elsewhere. Most of the companies that produced the home computer games were small firms, such as Sierra On-Line and Sirius, with sales of $10 million each.[14]

Games at that time were largely the work of programmers, either working on salary for the firms or freelance, sometimes for a share of the royalties. It was alleged that larger firms, such as Atari, gave little credit or publicity to these programmers who actually designed the games, refusing even to reveal their names in response to inquiries. Referred to as "towel designers," by the former textile industry managers who ran Atari, many of these "authors," as they were called in the industry at large, quit to go elsewhere or set up their own firms. "Atari," it was charged in 1984, "did not seem to address this loss outright, but instead focused its creative efforts on litigation and high-rolling licensing of seemingly fail-safe properties [like Pac-Man] from other media, from coin-operated games to movies." It seems probable that programmers at a place like Atari also had to periodically submit their "source codes" for new games being worked on so that the firm could exert greater supervision of development.[15]

Other programmers worked with more independence, like John Harris, a well-known pioneer in the field. Harris worked for what was then named On-Line Systems. Called both a "hacker" and a "nerd," Harris followed the traditional path of "hacker ethics," working partly for the "fun" of it and always chafing at any managerial control. On-Line hired him for a percentage of the future royalties of a Pac-Man clone, which he produced so faithfully that On-Line lawyers worried about copyright infringement. Sent back to rework the design, Harris rebelled, disguising his game by putting glasses and false noses (similar to the famous "Groucho-glasses") on the ghosts that threatened Pac-Man. Sent back again, he developed Jawbreakers, a game that would seem to have been sufficiently different to escape infringement. Atari sued, however, and the case was settled out of court.

One habit that marked Harris's independence, and frustrated company lawyers, was that he apparently did not work from a source code but directly from his head. He eventually went to work for Atari, converting the arcade game Frogger into one that could be used on the Atari home computer. It became the best seller of 1982.[16] Later some firms, as both a marketing strategy and perhaps to keep their outside authors happy, began to feature them as celebrities, providing, for example, photos of and interviews with the designers on the game packages. Individual programmers developed followings among gamers, who would develop loyalties to the "artist" rather than the company. "Among the star programmers," it was reported in 1983, "is Bill Budge, 28, who has had a cult following since he introduced *Raster Blaster*, a pin-ball simulation game" in 1981.[17]

The logical development of a video game, from the beginning, was (1) the conception of basic characters, background, sound effects, and play action; (2) the translation of this into a computer code, (3) the embodiment of the code in a ROM (read only memory) format; and, finally (4) the placement of the ROM into a computer and the latter into some kind of box. Where in this process the "authorship" of the game lay was not always clear.

The notion of giving legal protection to intellectual property is an old one. English patent practice dates from 1623 and by 1641 had been written into law in the American colony of Massachusetts. The American federal Constitution had authorized patents in 1787, and legislation in 1790 provided for the granting of patents for machines as well as copyrights for literary works. As technology expanded notions of publishing (phonograph records, films, xerography, and so forth), copyright protection was extended to them as well. The basic legal concept has remained from the beginning, however: ideas as such could not be copyrighted, only their concrete, distinctive, and specific *expressions*. The popular Pac-Man provided an illustration of this principle. In the case of *Atari, Inc. v. North American Phillips Consumer Electronics Corp.*, the court ruled that the idea of a figure in pursuit, a corral, and a maze pattern, all used in Pac-Man, were not protected by copyright since they were ideas.[18]

An argument in favor of such strict interpretation was that "progress in the arts . . . is a process of accretion through influence by prior works. Providing too much protection for very simple game themes and graphics might forestall the development of more complex works." Another legal problem for game developers was the tradition of requiring that, in order to be copyrighted, something had to be in a format that was readable by human beings. Thus, in the 1908 decision in *White-Smith Music Publishing Co. v. Apollo*, the court found that a piano roll of a copyrighted song was no infringement because the roll was a "part of a machine," not a direct form of communication.[19] Not surprisingly, patent litigation became a continuing reality in the video game industry. The programmer John Harris later recalled that "I was watching the computer game industry turn from a friendly and freestyle expression into a cutthroat business of large companies."[20]

Beginning in the late 1970s, Japan became a major center of video game technology. Significantly, many of the most important corporations involved had a history of producing coin-operated games for arcades. The game company Sega traced its roots back to a Honolulu company founded in 1940 to develop and sell coin-operated jukeboxes, games, and slot machines. It moved to Tokyo in 1951 and provided machines for the U.S. military bases in Japan. In the mid-'60s, it merged with another American company that imported coin-operated games, ran two-minute photo booths, and owned a chain of more than 200 arcades.[21] Taito had its start in 1954, importing and distributing vending machines in Japan. It then leased, and finally manufactured, jukeboxes and in the 1960s produced electromechanical games for arcades. Its first video arcade game was released in 1973, and five years later Toshihiro Nichikado, one of Taito's designers, created the very successful Space Invaders.[22]

Namco, a manufacturer of coin-operated games, began in Tokyo 1955, operating children's rides on the roofs of department stores in Yokohama, then in 1970 produced a coin-operated mechanical driving simulator called Racer. In 1974, Atari's Japanese branch was experiencing financial difficulties and Namco outbid

Sega for the company. At first Namco opened arcades featuring Atari games, but in 1978 introduced its first title, Bee Gee. The success of the arcade game Space Invaders inspired Namco to put Toru Iwatani, a former pinball machine designer, and his team to work on finding an equally successful arcade video game. The next year, Iwatani's Pac-Man was launched and proved to be an enormous success.[23] Namco sold 293,822 of the arcade machines between 1980 (when it was introduced into the United States) and 1987, while continuing to develop sequels such as Ms. Pac-Man, Pac-Mania 3D, and even a Pac-Man pinball machine.[24]

No Japanese company was more famous than Nintendo, which was founded in 1889 to make hanafuda playing cards by hand, later mass producing them and currently still manufacturing them. In 1966, it began to make toys, and in 1973 it took up family entertainment with the development of Laser Clay Shooting System venues using their own toy light guns in abandoned bowling alleys. Eight years later the company's head, Hiroshi Yamauchi, led the firm into the video game field, at first distributing the Magnavox Odyssey video game console in Japan. Then in 1977 Nintendo began to produce its own hardware. Yamauchi, however, who professed to know nothing about video games, claimed in 2004 that "cutting edge technologies and multiple functions do not necessarily lead to more fun. The excessively hardware-oriented way of thinking is totally wrong." Instead, he had a talent for recognizing good game designers.[25]

One such individual was Gunpei Yokoi, whose first success was a robotic arm that Nintendo used in games and gadgets. Another was Shigeru Miyamoto, who was asked to develop an arcade game and came up with Donkey Kong, a success in itself but also the origin of the Mario character. The Italian plumber with a mustache was a very popular character, and in a 2002 survey was recognized by more American children than was Mickey Mouse.[26] Reportedly, in 2013, eighteen of the twenty best-selling video games were made by Nintendo.[27]

While their games were very successful, Nintendo's hardware was also well received. Its first excursion into video games had been to distribute the Magnavox Odyssey video game console; then in 1983 it introduced its own Family Computer, a console known outside Japan as the Nintendo Entertainment System, or NES, whose main competitor was the Sega Mega Drive. Competition was also strong from Sony's PlayStation and the Microsoft Xbox, which Nintendo countered with their Wii, released in 2006. Meanwhile, in 1989, Nintendo introduced the Game Boy, which eventually became the best-selling handheld console on the market.[28]

Not all computer games were for home or arcade use; some built upon the proven success for more than half a century of mechanical gambling machines such as the slot machine, or "one-armed bandit." In 1885, Charles Fey arrived in California after having come to America from his native Bavaria by way of France and England. After first working for the California Electric Works, in 1894 he built his first slot machine and the next year his second, which he called the Liberty Bell. In 1896, he opened his own manufacturing firm to build the machines. The device itself consisted of a box containing three wheels with images of diamonds, spades, hearts, and a cracked Liberty Bell painted on them. After a nickel was deposited, the player pulled a handle (the bandit's one arm), spinning the wheels. If they came to a stop showing three of a kind, a prize of coins was dispensed. Fey refused to patent his machine or to sell them. Instead he leased them for 50 percent of the take.[29]

Gambling machines, like the widely popular Liberty Bell, were rapidly redesigned after World War II. In 1963 an electromechanical hopper, an internal payout device, was added, and with the advent of video games in the 1970s, reels and levers were replaced by screens and buttons, and in the 1980s microprocessors were added, most importantly the one that acts as a random number generator (RNG). International Game Technology, in 1982 headed by Si Redd, formerly of the pinball-making firm of Bally, became a success by marketing various types of video gambling games for Nevada casinos. More automated and reliable, they were reportedly taking a sizable share of the market from Bally's mechanical slot machines. "Just as *Pac-Man* replaced pinball in the arcades," reported *Forbes* magazine, "so, too, is video technology invading casinos." Redd was said to be not relying entirely on the tried and true gambling machines, but investing $4 million in research and development of new games.[30]

The experience of Nolan Bushnell (Pong) and Toru Iwatani (Pac-Man) with companies that manufactured pinball machines highlights the roots of the current close connection between video and gambling machines. The American firm WMS Gaming started up in 1974 as a maker of pinball machines, then switched in the 1990s to making slot machines for gambling. Reporting on WMS, one observer commented that slot machines were merely "video games that accept a wager." Laurie Lasseter, who earned her engineering degrees at Cornell University and served as the company's vice president for engineering and technology, asserted that the slot machines were following the "workstation trajectory," using off-the-shelf components. Computer games had become "the killer application for high-end processing power," and that proved "just fine with WMS."[31]

The resulting manipulation of "space and time to accelerate the extraction of money from players" has been most closely studied by the anthropologist Natasha

Dow Schull.[32] Critical to Schull's argument is the analogy between the gambler and Karl Marx's factory workers of the nineteenth century. As she points out, the object of the gambling machine designers to compress "a greater number of spending gestures into smaller units of time echoes Karl Marx's insight that 'moments are the element of profit,' along with Michel Foucault's apt characterization of modern disciplinary logic: 'it is a question of extracting, from time, ever more available moments and, from each moment, ever more useful forces.'" To help accomplish this, the designers add "a score of visual and auditory design elements—crisp, high-resolution graphics and enhanced animation as well as 'hi-fi' sound—[which] compose a 'second-order conditioning' that adds to the reinforcement of play."[33]

So popular have these video gambling machines become with the players that casinos have had to decrease the floor space used for traditional staff-mediated games, like roulette, and card games, like blackjack, played at a table with a dealer. The players, it seems, want to be alone and uninterrupted, playing as fast as possible. As one gambler told Schull, "The only thing that exists is the screen in front of you. You go into the screen, it just pulls you in." Players are said to enter "the zone," a kind of trance in which the object becomes not winning but continuing to play. "Once players are absorbed in the game," writes Schull, "the technological aesthetic features that draw them in stop mattering." She asserts that "as it turns out, at a certain point in the career of the digital game it became necessary to background the digital itself." "The intensification of digital capabilities," she concludes, "leads not only to an exit from embodied space and chronometric time, the dematerialization of money, and the cancellation of desire by way of its immediate fulfillment, but to the falling away of the material technology itself."[34]

In Australia, an AutoPlay option (not permitted in the United States) allows players to press a button and have the game play itself. Indeed, Schull claims that "the zone exemplifies traits of 'postmodernity' . . . : The zone is characterized by play rather than purpose, chance rather than design, absence and immersion rather than presence and perspective, the collapse of time and space (or 'time-space compression'), the 'waning of affect,' the free-floating circulation of credit in market exchange." It is ironic that the artistry of the game designers is most successful at the very point at which their work becomes invisible: "after the body folds into the machine," Schull writes, "and the machine folds into the game, what is left is the abstract, digital procedure of the game itself."[35]

In 2006, Renaud Donnedieu de Vabres, the French culture minister, presented the Ordre des Arts et des Lettres medal to three video game designers, including

Shigeru Miyamoto, the creator of Donkey Kong and the much-loved character Mario. "Call me," he said, "the minister of video games if you want—I am proud of this." His justification was his belief that "video games are not a mere commercial product. . . . They are a form of artistic expression involving creation from script writers, designers and directors." They may not have been "mere" commercial products, but they certainly were that in part, and France was home to a hundred game companies, including three of the world's top ten firms, among them Infogrames Entertainment which owned Atari. If the European Union accepted his designation, video games would be eligible for tax breaks of up to half a million euros.[36]

The question of whether or not video games are an art form is very much undecided, and opinions are mostly just that. One game designer is quoted as boasting that "games are not just an art. . . . They are the most revolutionary form of art mankind has ever known about." At the other extreme is the American critic Roger Ebert, who denies that video games can ever be art, for the structural reason that they "require player choices, which is the opposite of the strategy of serious film and literature, which requires authorial control."[37] It was "the nature of the medium" that "prevents it from moving beyond craftsmanship to the stature of art." This ignores, of course, the theory that it is readers, not writers, who always control the text, and there are many game designers who consider that very interactivity a major source of a game's artistic claim. Indeed, game advocates point to the additional fact that "not much difference can be observed between the methods of making games and animated films. . . . [Indeed] some of the artists and programmers cross between the industries."[38]

The debate has hardly risen to the level of serious aesthetic consideration, however. One writer for the journal *Aesthetics* lamented that "there has been no sustained argument on either side of the video games as art debate." Courts have not been able to avoid the question as easily as philosophers. In the case of *American Amusement v. Kendrick*, for example, it was argued that the freedom of speech guarantee of the U.S. Constitution should cover game contents because it shares themes with literature. The author reminds us that "Kantian aesthetics puts play as one of the central features of aesthetic experience," and suggests that perhaps one should "approach the medium's current state as similar to that of film in the late 19th century: we can see a continuum from the relatively primitive Luminere actualities such as 'Arrival of the Train' to the fully-realized promise of the artform that is obvious only decades later in the works of Fritz Lang."[39]

Thirty years after the first pings and pongs accompanied the earliest computer games, video game soundtracks had evolved into real music played by rock groups or symphony orchestras. Indeed, by 2007 it was claimed that "games are one of the most effective ways for musicians to establish and increase their audience."[40] A good half of American gamers, aged 13 to 32, claimed that they had discovered new bands from game soundtracks, often following up by downloading songs or buying the albums of the new groups. In part the new interest in game music came through the transition from using floppy disks and cartridge-based games to taking advantage of CDs and DVDs.

Not surprisingly the composer Hitoshi Sakimoto said that composing for games was much like composing for motion pictures, but at the other end of the scale, some gamers were producing their own soundtracks, taking tracks from their favorite CDs, mixing them, and then saving them to a console hard drive. A principal in one recording company declared that "soon these games will become the new radio, the new MTV and the new record store all in one."[41]

In a major attempt to translate game music to a concert venue for live audiences, Jason Michael Paul, the producer who had made such a success of the Three Tenors concerts, came up with the idea for *Play!—A Video Game Symphony* in 2004. With large-scale LCD screens showing scenes from the appropriate video games as well as close-ups of the orchestra's musicians, such orchestras as the Los Angeles Philharmonic, which put on the first show, the Stockholm Philharmonic, and the Sydney Symphony played two hours of game soundtracks, mixed with a movement from Gustav Holst's *Planets*. One cynical music critic at the Sydney concert remarked that "the trick . . . is to add an ironic edge to avoid bland pastiche."[42]

Any irony was probably lost on the crowd (mostly young, mostly male). The audience in Stockholm responded with "applause, some loud whoops and a stamping of feet," while a field of mobile phones were held high to record the moment. The enthusiastic, sell-out crowds were seized upon by some sanguine commentators as perhaps evidence that a younger generation had been given a positive introduction to the classical concert hall, and might return.[43] The composer of the music for the game World of Warcraft opined that "people are actually coming to recognize game music in its own right," and offered "a bridge to people who wouldn't normally set foot in a concert hall." The music critic was less enthusiastic. "Can junk culture aspire to art?" he asked, and answered "Of course it can."[44]

One measure of the way in which video games have made a claim to being artistic expressions is the way in which educational and professional structures have sprung up to define and nurture them. In 2002, the Barbican Art Gallery in London opened an exhibit called "Game On," which then moved to the Science Museum and thence around the world. In 2005, Stanford University offered an STS (Science, Technology and Society) course entitled "History of Computer Games Design: Technology, Culture, Business." In 2006, a website asked readers, "Are you ready to take it by the horns and ride? Perhaps video game design is the career for you. There really has never been a better time to get on board. Maybe you'll become the next hot *video game designer*." An online course, leading to a Bachelor of Science degree in Game Art & Design, was offered by the Art Institute of Pittsburgh, Pennsylvania, one of a number of schools to provide such education. "The Game Art & Design program," it promised, "concentrates on the artistic side of games—not computer programming. This unique program," it concluded, "is your first step toward becoming an artist and designer in the multi-billion dollar game design industry." In 2005, the Liverpool John Moores University and the School of Computing & Mathematical Sciences hosted the 3D international conference in Computer Game Design and Technology. Warwick University was home to the Warwick Video Game Design Society, designed to provide "a platform for students . . . to learn or practice skills in anything connected to multimedia electronic games—whether it be music, art, writing or technical proficiency in programming and 3D modeling (to give but a few examples)."[45]

Beyond the question of the current or potential merit of video games as art lies the complex web of institutions to which they are linked. Even a cursory review of their development since World War II reveals that, however novel they may appear to be in terms of popular culture, they share important connections with multiple strands of the history of modern technology in general. "Junk" may aspire to be art, but it remains firmly embedded in its technological origins and social networks.

The birth of video games in the nation's university laboratories during the Cold War meant that they were never far from the massive commitment to military research that dominated those years. One of the earliest games was called Hutspiel, built by the U.S. military in 1955 to model a war with the Soviet Union in Europe. Such games as Starwars appear almost benign in retrospect as compared with the "first-person shooters" that appeared as early as 1973, and in which the player is the armed perpetrator. The legislator who in 2005 authored a successful bill in California to limit the availability of violent games noted that "unlike movies where you passively watch violence, in a video game, you are the active participant and making decisions on who to stab, maim, burn or kill." The concern was

Violent video games, especially "first-person shooters," have been controversial. It has been argued by some that they stimulate and reinforce the violent urges of the young (usually men) who play them, but by others that they harmlessly discharge those same tendencies. With the permission of the artist, John Shakespeare (*Sydney Morning Herald*).

that "as a result, these games serve as learning tools that have a dramatic impact on our children."[46]

Not all "first-person shooters" were based on military combat, but the line between civilian and military violence could be a thin one. One U.S. marine, after a bloody engagement in an Iraqi town, was quoted by a journalist as exclaiming, "I was just thinking one thing when we drove into that ambush, *Grand Theft Auto: Vice City*. I felt like I was living it when I seen the flames coming out of windows, the blown-up car in the street, guys crawling around shooting at us. It was f——ing cool."[47] Prince Harry, the (as of 2013) fourth in line for the English throne, was condemned by the Taliban for expressing "joy" in his work of firing his Apache Hellfire helicopter's air-to-surface missiles, rockets, and 30mm machine gun. "It's a joy for me," he explained, "because I'm one of those people who loves playing PlayStation and Xbox, so with my thumbs I like to think I'm probably quite useful."[48]

The connection is neither casual nor coincidental. In its need to attract recruits, the U.S. Army has used such first-person shooters as America's Army, a "free download with online multiplayer capabilities" created specifically to appeal to gen(eration) Y players. Between 2002 and 2007, it attracted more than 8.6 million players, 28 percent of whom went on to click the link to the army's recruiting site. Forty percent of the recruits signing up in 2005 had previously played the game, which was also available for Xbox, mobile phone, and in coin-operated arcade versions.[49]

Ironically, the technology and efficacy of recruiting efforts through online games has proven attractive to jihadists as well as the U.S. military. According to a spokesperson for one video company, "Millions of people create mods on games around the world," and "we have absolutely no control over them. It's like drawing a mustache on a picture." The company's popular Battlefield 2 was one of the games modified by Islamists to show heroic urban guerillas fighting the crusading western aggressors. The presumed purpose was to "exhort Muslim youths to take up arms against the United States."[50] The technology of massive multiplayer online games, which allow people from around the world to team up and play in real time, has created, at least in the minds of security agencies, the possibility of terrorists hiding and plotting "in plain sight." It was reported in 2013 that spies from the FBI, the CIA, the Pentagon and England's Government Communications Headquarters had infiltrated the popular games World of Warcraft and Second Life. World of Warcraft had 12 million subscribers in 2010.[51]

When asked years later to describe his thinking when he designed the classic Pac-Man, Toru Iwatani replied that "all the computer games available at the time were of the violent type—war games and space invader types. There were no games that everyone could enjoy, and especially none for women. I wanted to come up with a 'comical' game women could enjoy." The game, based on eating rather than killing, was indeed popular with women, as it was, of course, with men and children as well.[52] Iwatani had identified one of the most persistent assumptions in the gaming world: that girls don't like violent games and want an entirely different kind. It was, of course, not only stereotypical but also circular in its cause and effects. Game makers directed their efforts at the perceived market, which was overwhelmingly male. It was overwhelmingly male, in turn, because violent action games appealed them. At the same time, not only are most gamers male but so are those who write the games, as are the featured characters.

The strongly gendered nature of video games makes them no different than any other type of game, from Erector sets to dolls. But just as some girls have always played with the one and some boys with the other, in 1994 it was reported that 21 percent of video games were purchased and used by women, and in 2012, 42 percent of gamers online were female. It turns out that the "prepubescent males" and the "men trying to entertain their inner boy" were no longer alone.[53] Anecdotal evidence, however, suggests that those females who do go against the grain sometimes pay a heavy price.

One 23-year-old female student who reported playing Call of Duty: Modern Warfare online nearly every night was the target of extreme misogyny, one night being told "f——g dumb bitch, I hope a f——g n——r rapes you and f——g kills you and your family." In less graphic language she is often told that, being a woman, she has no right to play video games. Some women hide their sex by adopting a man's name, play as male characters, or refuse to speak so that their voices do not betray them. Such abuse is often excused as the normal "trash talk" that is appropriate to the aggression of the game itself, but it is abuse that is pointedly directed toward the offending player's sex.[54]

Another hurdle for the female gamer is the dearth of female characters, especially in lead roles. Going back at least as far as Donkey Kong, the damsel-in-distress role was most common. More recently some heroines with active, even adventuress, roles have emerged, but even so, as one critic observed, "from the day when video games developed the graphic capabilities to simulate cleavage, women in games have been designed mostly from the blueprints of a horny 13-year-old boy's fantasy." The most recognized game character is in fact Lara Croft, the star of the 1996 game Tomb Raider. Croft's creator, Toby Gard, wanted her to be in the Indiana Jones tradition: "Lara," he said, "is fire and ice, mind and passion, strength and agility, but she is also an unusual anti-hero." He quit after the first game was released because Croft's breasts had been drastically inflated. "Over the years," he complained, "she was marketed in more and more sleazy ways, which were completely contrary to the goal of the character." In 2012 it was announced that in the next sequel, Croft would be the victim of an attempted rape by a gang of scavengers. Despite this, Croft had a following among women game players, presumably at least in part because she was a rare female role model.[55]

One major reaction to the gendered nature of video games was the attempt to create titles that would appeal more to girls. One scholar has referred to this as the " 'girls' games' movement." Propelled by both entrepreneurs in the industry who wanted to tap into a potentially lucrative new market and feminists who wanted to end the gendering of digital technology, conflicting aims sometimes led to less than successful results.[56] Whereas the latter tended to favor good games that would appeal to both boys and girls or games that had strong, proactive female characters, the former seemed to be locked into producing gender-stereotyped "girlie" games featuring shopping, dressing up, and gossip—all wrapped in pink.

One game, which came on five CD-ROMs, was called McKenzie & Co., described by one observer as "unmistakably girlish." The action is described as "classmates share advice, commiserate on failure and rely on each other to propel them into romance. And nerve-wracking conversations with desired males can be replayed with different girlish responses, teaching, again and again, that kindness, consideration and charm work far more effectively than cranky aggression."[57] The best known of this genre was Barbie's Fashion Designer, which came in a pink box and allowed users to create up to 15,000 different outfits and then walk Barbie herself down a runway in 3D.[58]

The culture of games was even more white than it was male. The best-known, and possibly only, African American central character in a major title is C.J., the violent criminal in Grand Theft Auto: San Andreas, who specializes in stealing cars and drive-by shootings. A 2005 survey found that 80 percent of video game programmers were white, 4 percent Hispanic, and less than 3 percent black. Like the effort to involve more women in creating games, programs were set up to encourage more participation from racial minorities. As one game design professor at the Georgia Institute of Technology observed, "For a long time, we've talked in the game industry about gender diversity as the one problem on the radar, but the racial split is worse."[59]

The issues raised in video game culture over race, gender, and violence are not unique, of course, and merely serve to demonstrate how deeply embedded the games are in American culture more generally. They share that condition with more traditional toys, as well as sharing the rich mix of production and consumption, entrepreneurial ambition, and artistic application. Despite the cutting-edge, high-tech glamour of video games, they are in many ways only a recent iteration of long established forces in American society.

Eight Hours for Recreation

R obert Owens, the philanthropic mill-owner of early nineteenth-century England, is credited with coining the phrase "eight hours labour, eight hours rest, eight hours recreation" to describe the ideal division of the day for the working man. In the latter years of the century it became a motto of the crusade for the Eight Hour Day, a key demand of the growing union movement in Anglophone countries. It was a radical ideal, a bold demand, and eventually a successful reform. Not surprisingly, social reality was a bit more complex. For one thing, the "working man" might be able to spend his time between work and sleep in recreation, but working women almost always had to work a second shift—cooking, cleaning, and taking care of the kids. For her, recreation was not so easily claimed. For another thing, children (at least before they were sent out to work) often had lives that consisted of nothing but rest and recreation, at least until school intervened.

When men, women, and children all found time for (usually homosocial) rec-reation, much of it was taken up with one or another form of play. And ironically, as the nineteenth century wore on, the distinction between their work and play became problematic in several respects. By the end of the century, for example, crowds of working-class people, of both sexes, were enjoying the new amusement parks where mechanical rides reproduced a form of the technological environ-ments in which they habitually worked and traveled. The middle class was being exhorted to spend their leisure time "working" at hobbies that reinforced the skills and motives of "real" work.

And children, as always, were given toys that were considered gender appropriate and that ideally would help prepare them for the working world of adults. Play was children's work, it was said, and toys were their tools. A prime example of this, of course, was giving dolls to girls and construction and science kits to boys to prepare them for life in an industrial world. Even playgrounds were recruited in an attempt to get working-class children used to the disciplines of industrial work. Organized sports became increasingly routinized, with teams and leagues, rules and regulations, and strict specifications governing the equipment to be used. While members of teams "played" baseball they were, of course, professionals actually working at their craft. A striking number of people with engineering training seem to have taken up, on a temporary or part-time basis, the design and development of sailboards, snowboards, and hang gliders.

Another common phenomenon was the flow of ideas, technologies, and individual entrepreneurs and inventors between the various forms and venues of play. One striking example was A. C. Gilbert, who is best known for his invention of the construction toy Erector. His Erector sets were advertised as being used to make Ferris Wheels and other rides made famous by amusement parks, as well as skyscrapers, and various forms of transportation such as wagons and wheelbarrows, weaving looms, and all manner of other devices familiar to the worlds of work and play. In the context of World War I, Gilbert marketed toy machine guns, submarines, and nurses' kits. At the same time he ventured into other forms of play, and patented a vibrator and a hand-held electric drill for do-it-yourselfers. Lego began as a toy and then became a video game and even a theme park.

The engineer Frederick W. Taylor, putative Father of Scientific Management and initiator of the efficiency craze in the early twentieth century, turned his attention to the sports he loved to play and took out patents for both an "improved" tennis racket and a new way to rig a tennis net. He was also an avid golfer and designed, and used, a radically re-designed putter as well as conducted a running research project on the best turf for golf courses. Indirectly he was also an influence in the rational playground movement. In baseball, Albert Spalding went from being a star player for the Chicago White Stockings to playing a leading role in promoting the sport of baseball, manufacturing baseball equipment, and codifying the rules and records of the game. The architect Frank Lloyd Wright credited a childhood love of building blocks as an influence on his career, and his son patented the successful and long-lived construction toy Lincoln Logs.

The distinction between toys and "real" tools was sometimes difficult to define. Bicycles were a prime example, used at the same time for adult transportation, in a professional sport, and as a toy for children. The BB gun was sold as a toy but was also used to initiate children into a formal gun culture with military overtones. Carpentry tools, usually organized into boxed sets, were given to boys as toys, although they were identical to those used in the building trades, even if usually scaled down in size.

While being embedded in the modern world of industry and invention, the technologies of play, along with the venues in which they were used, were also deeply implicated in the national cultures of racial and gender discrimination. Amusement parks were often segregated and African Americans were subjected to the humiliation of side-show exploitation. Parks built and patronized by black Americans were one response to this discrimination. Major Taylor was an internationally acclaimed bicycle racing champion at the turn of the twentieth century, but became lost to the history of the sport. In baseball, the Negro League flourished, but outside the mainstream of the national sport.

Women and girls were always treated differently, that is to say, not as equals to their male counterparts. They played with the same toys, participated in the same sports, became involved in the same hobbies, but this participation was never officially or adequately acknowledged. Girls were encouraged to play with the "appropriate" toys, and were grudgingly allowed to organize segregated teams for sports. The appearance of extreme sports, with its aggressively masculinist culture, opened a new arena for discrimination. Again girls and women surfed, and rode skateboards and snowboards, but were always seen as not only outside the mainstream of the sports, but worse, threatening to erode the hypermasculine culture and value of the activities. The technologies might be the same, but the cultural significance was vastly different. Significantly, even as some traditional nineteenth-century venues for male homosocial leisure, such as saloons and fraternal orders, began to fade in the twentieth century, something of their role was taken over by hobbies such as ham radio and car modification, which allowed men to separate themselves from their families and commune with other like-minded (male) enthusiasts.

The free choice of play had the potential to destabilize the gender system of the broader society. Frances Willard was forced to abandon her childhood playthings, and along with them a freedom from gender constraints which she was not able to recapture until she discovered the bicycle, a technology that had begun as a masculine preserve and only later was captured by women as a liberating technology for them as well. Many a girl played with an Erector set, but almost always it was one that had been purchased for her brother. Most recently the computer, the Web, Internet, and video games have created a new arena for gender conflict, and a venue for a shocking degree of misogyny.

As was true with other aspects of American life, the technologies of play and their meanings intertwined and evolved in an intricate matrix of reciprocity. Historians realize that social constructions, from gender and racial systems to sports and amusement parks and the modern corporation, are always contested and contingent. Play, in all its manifestations, has been no different. There have always been large economic advantages and important cultural values in play, and we are all implicated, however unequally, in their sorting out.

Notes

Introduction: Playing with Technology

1. Gary Cross, "Play in America from Pilgrims and Patriots to Kid Jocks and Joystick Jockeys: Or How Play Mirrors Social Change," *American Journal of Play* (Summer 2008), 8.

Chapter One: Toys for Girls and Boys

Epigraphs. A. C. Gilbert Co., *Boy Engineering* (New Haven, 1920), p. 105. Christine Frederick, "Grown-up Accessories for Small People," *American Home*, 1 (December 1928), 227.

1. Frances E. Willard, *How I Learned to Ride the Bicycle* (Sunnyvale: Fair Oaks Publishing Co., 1991 [1895]), pp. 1–2, 76.

2. This description of the doll trade relies heavily on Miriam Formanek-Brunell, *Made to Play House: Dolls and the Commercialization of American Girlhood, 1830–1930* (New Haven: Yale University Press, 1993).

3. U.S. Patent No. 681,974, issued September 3, 1901.

4. Reproduced in Formanek-Brunell, *Made to Play House*, p. 36.

5. U.S. Patent No. 1,382,708 issued to David Zaiden, June 28, 1921.

6. Judy Attfield, "Barbie and Action Man: Adult Toys for Girls and Boys, 1959–93," in *The Gendered Object*, ed. Pat Kirkham (Manchester: Manchester University Press, 1996), pp. 80–89.

7. Tonka Corp., *Tonka '73* (n.d.), p. 34.

8. Donald W. Ball, "Toward a Sociology of Toys: Inanimate Objects, Socialization, and the Demography of the Doll World," *Sociological Quarterly*, 8 (1967), 447.

9. Louis Wolf Goodman and Janet Lever, "Children's Toys and Socialization to Sex Roles" (August 1972), unpublished, quoted by permission.

10. See, for example, W. Ogden Coleman, "Toys, Moulders of Industrialists," *American Industries*, 22 (August 1921), 13.

11. Harry Jerome, *Mechanization in Industry* (New York: National Bureau of Economic Research, 1934), p. 432.

12. Christine Frederick, "Grown-up Accessories for Small People," *American Home*, 1 (December 1928), 227, 276, 278.

13. Kathryn Tuggle, "Hasbro Cooks Up Something New: A Bulb-less Easy-Bake Oven," Fox News, September 2011, http://www.foxbusiness.com/personal-finance/2011/09/14/hasbro-cooks-up-something-new-bulb-less-easy-bake-oven/. Accessed September 15, 2011.

14. Ad for Stanlo in *Parents Magazine*, 8 (December 1933), 65.

15. A. C. Gilbert Co., *Boy Engineering* (New Haven, 1920), p. 105.

16. Edwin T. Hamilton, "Gifts for Them to Make," *Parents Magazine*, 9 (December 1934), 81.

17. George Carl Weller, "Toymaking for Girls," *Education*, 53 (October 1932), 112–113.

18. U.S. Patent No. 1,475,575 granted November 27, 1923.

19. U.S. Patent No. 54,207 granted November 18, 1919.

20. U.S. Patent No. 1,457,972 granted June 5, 1923.

21. "Radio Flyer," http://en.wikipedia.org/wiki/Radio_Flyer. Accessed December 7, 2011. (Cleveland paper) *The Plain Dealer*, May 3, 1997.

22. Dail Steel Products Co., *Wolverine Juvenile Vehicles* (Lansing, 1924), p. 15.

23. Auto-Wheel Coaster Co., *The New Models* (North Tonawanda, 1921), p. 15.

24. Gilbert Co., *Boy Engineering*, p. 121.

25. Ibid., p. 76.

26. Frederick Lowey & Co., *Lowey's First Steps in Chemistry* (New York, 1882).

27. Porter Chemical Co., *Chemcraft No. 10 Experiment Book* (Hagerstown, 1928), p. 11.

28. Ibid.

29. Alfred Albelli, "Toys Make the Man," *Popular Mechanics*, 52 (December 1929), 962–963.

30. Morris Meister, "The Educational Value of Scientific Toys," *School Science and Mathematics*, 22 (December 1922), 801–813.

31. Obituary notices of Hornby in *New York Times*, September 22, 1936, and *The Times*, September 22, 1936.

32. Meccano Co., *Meccano Prize Models: A Selection of the Models Which Were Awarded Prizes in the Meccano Competition 1914–15* (New York: Duell, Sloan and Pearce, 1915), p. 13.

33. Frank Lloyd Wright, *An Autobiography* (New York: Duell, Sloan and Pearce, 1943).

34. "Milton Bradley Company," http://en.wikipedia.org/wiki/Milton_Bradley_Company.html. Accessed January 19, 2012.

35. Erin K. Cho, "Lincoln Logs: Toying with the Frontier Myth," *History Today*, 43 (April 1993), 31.

36. U.S. Patent No. 1,351,086 granted August 31, 1920.

37. G. A. Nichols, "How Advertising Opened All-Year Market for Toys," *Printer's Ink*, 123 (April 26, 1923), 124.

38. President, The Toy Tinkers, Inc., "Manufacturing Policies That Have Offset Europe's Cheap Labor," *Factory*, 33 (December 1924), 779–781, 846.

39. Quoted in his obituary notice, *New York Times*, January 25, 1961.

40. Gilbert Co., *Boy Engineering*, p. 3.

41. Maaike Lauwaert, "Playing Outside the Box—on LEGO Toys and the Changing World of Construction Play," *History and Technology*, 24 (September 2008), 223.

42. Ibid., 225, 226.

43. "LEGOS," http://web.mit.edu/invent/iow/christianse.html. Accessed December 8, 2011.

44. "Man Builds a Living Out of LEGO," http://edition.cnn.com/2007/SHOWBIZ/

05/31/lego.artist/index.html?iref=allsearch. Accessed June 1, 2007, *Sydney Morning Herald*, January 27, 2008.

45. http://mindstorms.lego.com/en-us/history/default.aspx.

46. Dave Baum et al., *Extreme MINDSTORMS: An Advanced Guide to LEGO MIND-STORMS* (Berkeley: Apress, 2000), p. 3.

47. *Sydney Morning Herald*, December 22, 2011.

48. *Parents Magazine*, 8 (December 1933), 2.

49. "Toy Weapon," http://en.wikipedia.org/wiki/Toy_weapon. Accessed November 24, 2011.

50. Rogers Daisy Airgun Museum, http://www.daisymuseum.com/html/timeline/1970.html. Accessed November 24, 2011.

51. Rogers Daisy Airgun Museum, http://www.daisymuseum.com/html/timeline/1890.html. Accessed November 24, 2011.

52. Rogers Daisy Airgun Museum, http://daisymuseum.com/html/timeline/1910.html. Accessed November 24, 2011.

53. Rogers Daisy Airgun Museum, http://www.daisymuseum.com/html/timeline/1930.html. Accessed November 24, 2011; and http://www.daisymuseum.com/html/timeline/1940.html. Accessed November 24, 2011.

54. Rogers Daisy Airgun Museum, http://www.daisymuseum.com/html/timeline/1960.html. Accessed November 24, 2011.

55. U.S. Patent No. 1,351,565 granted August 31, 1920.

56. U.S. Patent No. 1,310,613 granted July 22, 1919.

57. Ibid., p. 120.

58. A. C. Gilbert and Marshall McClintock, *The Man Who Lives in Paradise* (New York: Heimburger House, 1954), p. 159.

59. Sake, *The Short Stories of Saki (H.H. Munro)* (New York: Modern Library, 1958), pp. 441–446.

60. http://www.ecocitybuilders.org/richard-register/personal-oddesy/early-years-no-war-toys-era/. Accessed November 24, 2011; Richard Register, "War Toys and Ecocities," http://www.ecotecture.com/library_eco/editorials/ed_wartoys_rr.html. Accessed November 24, 2011. Also see Carol Andress, "War Toys and the Peace Movement," *Journal of Social Issues*, 25 (January 1969), 85–100.

61. Coleman, "Toys, Moulders of Industrialists," p. 13.

62. Gilbert and McClintock, *The Man Who Lives in Paradise*, pp. 158–159, 144.

63. Goodman and Lever, "Children's Toys and Socialization to Sex Roles."

64. Zoe Williams, "Hamleys' Baby Steps Toward Gender Equality," *Guardian*, December 13, 2011.

65. Kira Cochrane, "The Fightback Against Gendered Toys," *Guardian*, April 23, 2014.

66. Ibid.

67. David Crouch, "Toys R Us's Stockholm Superstore Goes Gender Neutral," *Guardian*, December 24, 2013.

68. Henry Petroski, "Work and Play," *American Scientist*, 87 (May–June 1999), 208, 209.

69. Ibid., 211.

70. Frank Lloyd Wright, *An Autobiography*, p. 13.

71. *Sydney Morning Herald*, May 27–28, 2006.

72. "Martin L. Perl—Autobiography," Nobelprize.org; http://nobleprize.org/physics/laureates/1995/perl-autobio.html. Accessed September 8, 2005.

73. "In Depth: National Engineers Week," *Business Journal*, February 26, 1999, http://bizjournals.com/sanjose/stories/1999/03/01/focus3. html. Accessed September 8, 2005.

74. Judy Jackson, "All I Want for Christmas Is . . . a Hydrodynamics Kit? *F.E.R.M.I.*, 22 (December 17, 1999), http://final.gov/pub/ferminews/ferminews99–12–17/p1.html. Accessed September 8, 2005.

75. Sir Harold W. Kroto, "Autobiography," www.nobelprize.org/nobel_prizes/chemistry/laureates/1996/kroto.html. Accessed January 25, 2012.

76. Eugene S. Ferguson, *Engineering and the Mind's Eye* (Cambridge: MIT Press, 1992), p. 42.

77. Ibid., p. 58.

78. Ibid., pp. 57, 166.

79. "Toy Models Help to Solve Many Problems for the Engineer and Inventor," *Popular Mechanics*, 40 (December 1923), 934.

80. Doris Bergen, "Play as the Learning Medium for Future Scientists, Mathematicians, and Engineers," *American Journal of Play* (Spring 2009), 413, 416, 420, 425.

81. Mark Dodgson, David Gann, and Ammon Salter, *Think, Play, Do: Technology, Innovation, and Organization* (Oxford: Oxford University Press, 2005), pp. 117, 108, 120–121.

82. Tracey Schelmetic, "Where Are America's Women Engineers?" February 19, 2013, news.thomasnet.com/imt/2013/02/19/where-are-america's-women-engineers/. Accessed April 29, 2014.

83. http://www.goldieblox.com/pages/about. Accessed April 27, 2014.

Chapter Two: The Safe and Rational Playground

Epigraphs. International Play Association, "Global Consultations on Children's Right to Play," http://article31.ipaworld.org/. Accessed December 1, 2011. J. Huizinga, *Homo Ludens: A Study of the Play-Element in* Culture (London: Routledge & Kegan Paul, 1949), p. 7.

1. Allen Guttmann, "The Progressive Era Appropriation of Children's Play," *Journal of the History of Childhood and Youth*, 3 (Spring 2010), 147.

2. See Cary Goodman, *Choosing Sides: Playground and Street Life on the Lower East Side* (New York: Schocken Books, 1979).

3. Ibid.

4. David Nasaw, *Children of the City: At Work & At Play* (New York: Anchor Press/Doubleday, 1985), p. 30.

5. Quoted in ibid., pp. 20–21.

6. Peter and Iona Opie, *Children's Games in Street and Playground* (Oxford: Oxford University Press, 1970), p. 20.

7. *New York Times*, May 6, 1910.

8. Sadie American, "The Movement for Small Playgrounds," *American Journal of Sociology*, 4 (1898), 165.

9. Rebecca Mead, "State of Play," *New Yorker*, 86 (July 5, 2010), 32.

10. American, "The Movement for Small Playgrounds," 159–161.

11. Ibid., 162–163.

12. "Playgrounds and Small Parks" and "Playground Movement," in *Encyclopedia of Chicago*, http://www.encyclopedia.chicagohistory.org.

13. Charles Zueblin, "Municipal Playgrounds in Chicago," *American Journal of Sociology*, 4 (1898), 151, 154, 158.

14. *Los Angeles Times*, September 14, 1904.

15. *Los Angeles Times*, May 19, 1905.

16. Nasaw, *Children of the City*, 36.

17. *New York Times*, April 28, 1910.

18. The following section on Scientific Management draws heavily upon William Brian Mitchell, " 'We Want a Play Factory'. Scientific Management and Urban Playgrounds 1907–1929: The Example of Cleveland" (Master's thesis, History, Case Western Reserve University, May 2000). I am grateful for his help.

19. This same year a Playground Association, also guided by scientific research and aimed both at producing healthier, stronger children and stamping out larrikinism, began opening playgrounds in Sydney. Jan Kociumbas, *Australian Childhood: A History* (St. Leonard's: Allen & Unwin, 1997), pp. 161–162.

20. *Washington Post*, April 13, 1906.

21. Some brief biographical information on Gulick is provided in Thomas Winter, "Luther Halsey Gulick," *Encyclopedia of Informal Education*, www.infed.org/thinkers/gulick.htm.

22. The Gilbreths filmed workers, broke down their motions into "therbligs" (Gilbreth spelled backwards), then rearranged them into more "efficient" patterns of work.

23. *New York Times*, April 28, 1961.

24. Mitchell, "We Want a Play Factory," p. 19.

25. Ibid.

26. Quoted in ibid., pp. 21–22.

27. Quoted in ibid., pp. 23–24.

28. Ibid., pp. 25–27.

29. Roy Rosenzweig, *Eight Hours for What We Will: Workers and Leisure in an Industrial City, 1870–1920* (Cambridge: Cambridge University Press, 1983), p. 132.

30. Ibid., pp. 140, 143.

31. Ibid., p. 145.

32. Ibid., pp. 146–147, 150–151.

33. Mitchell, "We Want a Play Factory," pp. 28–30, 36–45.

34. Frederick Winslow Taylor, *The Principles of Scientific Management* (New York: Harper & Brothers, 1911), p. 7.

35. *New York Times*, January 5, 1908.

36. *Washington Post*, April 15, 1906.

37. See, for example, Ann Taylor Allen, "Gender, Professionalization, and the Child in the Progressive Era: Patty Smith Hill, 1868–1946," *Journal of Women's History*, 23 (Summer 2011), 112–136.

38. Obituary of Carmelita Hinton in the *New York Times*, January 23, 1983; "Carmelita Hinton," http://en.wikipedia.org/wiki/Carmelita_Hinton; quoted in "Jungle gym," http://en.wikipedia.org/wiki/ Jungle_gym. U.S. Patent No. 1,488,245 for a Climbing Structure, application filed October 1, 1920, granted March 25, 1924.

39. *New York Times*, March 30, 1925; July 15, 2000.

40. *New York Times*, August 7, 1927.

41. *New York Times*, September 8, 1933.

42. *New York Times*, April 5, 1936.

43. *New York Times*, November 11, 1936.

44. Quoted in Leslie McGuire, "Isamu Noguchi's Playground Designs," Landscape-Online.com, 2010, http://www.landscapeonline.com/research article/7115. Roy Kozlovsky, "Adventure Playgrounds and Postwar Reconstruction," in *Designing Modern Childhood. History, Space, and the Material Culture of Children: An International Reader*, ed. Marta Gutman and Ning de Conick-Smith (New Brunswick: Rutgers University Press, 2007), p. 188.

45. Quoted in McGuire, "Isamu Noguchi's Playground Designs."

46. Susan G. Solomon, *American Playgrounds: Revitalizing Community Space* (Hanover: University Press of New England, 2005), p. 13.

47. Lianne Verstrate and Lia Karsten, "The Creation of Play Spaces in Twentieth-Century Amsterdam: From an Intervention of Civil Actors to a Public Policy," *Landscape Research*, 36 (February 2011), 85–109; Lady Allen of Hurtwood, *Planning for Play* (London: Thames and Hudson, 1968).

48. *New York Times*, October 7, 1951.

49. Quoted in McGuire, "Isamu Noguchi's Playground Designs."

50. Quoted in *New York Times*, March 25, 1952.

51. *New York Times*, August 3, 1953.

52. Solomon, *American Playgrounds*, p. 29; *New York Times*, July 4, 1954.

53. Solomon, *American Playgrounds*, p. 30.

54. Ibid., p. 27.

55. *New York Times*, October 3, 1953.

56. Thomas Meehan, "Creative (and mostly upper-middle-class) Playthings," *Saturday Review*, 55 (December 16, 1972), 45.

57. For the exhibit itself, see Karen Ann Marling, *As Seen on TV: The Visual Culture of Everyday Life in the 1950s* (Cambridge: Harvard University Press, 1994), chapter 7, "Nixon in Moscow: Appliances, Affluence, and Americanism," pp. 242–283.

58. *New York Times*, November 6, 1966.

59. *New York Times*, May 19, 1967. See also Richard Dattner, *Design for Play* (Cambridge: MIT Press, 1974).

60. U.S. Patent No. 3,632,109, filed July 22, 1969, granted January 4, 1972.

61. Response by Dattner to blog "daddytypes.com," March 25, 2009, daddytypes.com/2009/03/25/Richard_dattner_habitot_the_ur-playcubes.pph.

62. McGuire, "Isamu Noguchi's Playgrounds;" *New York Times*, October 24, 1962; February 5, 1964; October 6, 1966.

63. *New York Times*, December 9, 1962.

64. Piedmont Park Conservancy, 2009, "Playgrounds," http://www.piedmontpark.org/do/playgrounds.html.

65. *Sydney Morning Herald*, June 8, 2007.

66. Mead, "State of Play," 33–34, 36.

67. http://www.miracle-recreation.com/about-us.

68. Solomon, *American Playgrounds*, p. 82.

69. *New York Times*, May 20, 1999.

70. *New York Times*, July 4, 1954.

71. *New York Times*, May 16, 1965.

72. Solomon, *American Playgrounds*, p. 78.

73. Ibid., p. 79.

74. U.S. PIRG Reports, "Playing It Safe: The Sixth Nationwide Safety Survey of Public Playgrounds," June 20, 2002, http://static.uspirg.org. PIRG and the Consumer Federation of America had reported similar findings in 1994. *New York Times*, May 29, 1994.

75. USA TODAY, "The Great American Swing Set Is Teetering," November 25, 2008, http://usatoday.com.

76. Tim Gill, *No Fear: Growing Up in a Risk Adverse Society* (London: Calpuste Gulbenkian Foundation, 2007), p. 10.

77. Reyhand Harmanci, "The Bay Area's Best Outdoor Playgrounds," *San Francisco Chronicle*, March 15, 2009.

78. See Ronald D. Cohen, "Child-Saving and Progressivism, 1885–1915," *American Childhood: A Research Guide and Historical Handbook*, ed. Joseph M. Hawes and N. Ray Hiner (Westport: Greenwood Press, 1985), pp. 273–309.

79. Nicholas Sammond, *Babes in Tomorrowland: Walt Disney and the Making of the American Child, 1930–1960* (Durham: Duke University Press, 2005), p. 6.

80. Huizinga, *Homo Ludens*, p. 10.

Chapter Three: From Pleasure Gardens to Fun Factories

Epigraph. Quoted in "Amusement Parks," *Cleveland History*, http://ech.case.edu/ech-cgi/article.pl?id=AP1. Accessed January 29, 2009.

1. *San Francisco Chronicle*, July 23, 2010; U.S. Patent No. 3,995,460 issued December 7, 1976.

2. Kathy Peiss, *Cheap Amusements: Working Women and Leisure in Turn-of-the-Century New York* (Philadelphia: Temple University Press, 1986), p. 7.

3. Michael Sorkin, "See You in Disneyland," in *Variations on a Theme Park: The New American City and the End of Public Space*, ed. Michael Sorkin (New York: Noonday Press, 1992), p. 223.

4. Sarah Jane Downing, "Green and Pleasant," *Observer*, April 5, 2009.

5. "Vauxhall Gardens 1661–1859. Brief History," http://www.vauxhallgardens.com/vauxhall_gardens_briefhistory_page.html. Accessed February 15, 2011.

6. http://www.tivoli.dk/composite-4609.htm. Accessed October 7, 2011.

7. Carroll Pursell, *The Machine in America: A Social History of Technology*, 2d ed. (Baltimore: Johns Hopkins University Press, 2007), pp. 135–136.

8. Day Allen Willey, "The Trolley-Park," *The Cosmopolitan*, 33 (July 1902), 265.

9. Ibid., 266.

10. Ibid., 267, 269.

11. Quoted in "Carousel," *OED* online.

12. Frederick Fried, "Last Ride for the Carousel Figures? *Historic Preservation*, 29 (July–September 1977), 22.

13. "The Dentzel Family Carousel Story," http://dentzel.com/coloringbook/story/. Accessed October 8, 2011.

14. Ibid.

15. From *A Standard History of Kansas and Kansans*, written and compiled by William E. Connelley, published in 1918. http://www.skyways.org/genweb/archives/1918ks/biop/parketcw.html. Accessed October 9, 2011.

16. Fried, "Last Ride," 26.

17. University of Sheffield, National Fairground Archive, http://www.nfa.dept.shef.ac.uk/history/rides/history.html. Accessed October 3, 2011.

18. National Park Service, "Dentzel Carousel Restoration," http://www.nps.gov/partnerships/dentzel_carousel.html. Accessed October 8, 2011.

19. Peter Mundy, *The Travels of Peter Mundy, in Europe and Asia, 1608–1667*, The Hakluyt Society, 2d series, No. 17 (Cambridge: Hakluyut Society, 1907), p. 59.

20. Norman D. Anderson and Walter R. Brown, *Ferris Wheels* (New York: Pantheon Books, 1983), pp. 7–13; "Flying Horses Carousels," http://en.wikipedia.org/wiki/Flying_Horses_Carousel. Accessed November 21, 2011.

21. U.S. Patent No. 489,238 granted January 3, 1893; "Ferris Wheel," http://en.wikipedia.org/wiki/Ferris_wheel. Accessed November 22, 2011.

22. Anderson and Brown, *Ferris Wheels*, pp. 16–30.

23. William H. Searles, "The Ferris Wheel," *Journal of the Association of Engineering Societies*, 12 (November 1893), 619.

24. Anderson and Brown, *Ferris Wheels*, pp. 31–35.

25. Ibid., pp. 36–39; Eli Bridge Company, http://www.elibridge.com/history.html. Accessed November 21, 2011.

26. U.S. Patent No. 796,772 granted August 8, 1905.

27. Anderson and Brown, *Ferris Wheels*, p. 43; "Ferris Wheel," http://en.wikipedia.org/wiki/Ferris_wheel. Accessed November 22, 2011; *San Francisco Chronicle*, September 10, 2013.

28. U.S. Patent No. 54,202 granted November 18, 1919; U.S. Patent No. 1,935,558 granted November 14, 1933.

29. Adam Sandy, "Roller Coaster History: How It Started," http://www.ultimaterollercoaster.com/coasters/history/start/. Accessed November 17, 2011.

30. U.S. Patent No. 310,966 granted January 20, 1885.

31. Arwen P. Mohun, "Designed for Thrills and Safety: Amusement Parks and the Commodification of Risk, 1880–1929," *Journal of Design History*, 14, no. 4 (2001), 291–306.

32. William J. Lampton, "The Fascination of Fast Motion," *The Cosmopolitan*, 33 (June 1902), 123, 126.

33. "Roller Coaster History: Ride Designers," http://www.ultimaterollercoaster.com/coasters/history/designer/designer.shtml. Accessed November 17, 2011; David Pescovitz, "Roller Coaster Biographies: John Allen 1907–1979," http://search.eb.com/coasters/i_allen.html. Accessed May 30, 2005.

34. "John Miller," Lemelson-MIT, http://web.mit.edu/invent/iow/miller.html; David Pescovitz, "Roller Coaster Biographies: John Miller 1872–1941, http:www.britannica.com/coasters/i_miller.html; "Philadelphia Toboggan Company Inc., http:www.philadelphiatopboggancoastersinc.com/history.php.

35. "Tail Spinning," *Sunset* (July 1976), 70; "Tallest, Fastest Roller Coaster on Earth Debuts," http://www.ultimaterollercoaster.com/news/stories/20050519_01.shtm. Accessed May 30, 2005.

36. U.S. Patent No. 421,887, applied for in 1899, granted February 25, 1890.

37. U.S. Patent No. 1,339,299 granted May 4, 1920.

38. U.S. Patent No. 1,339,299.

39. U.S. Patent No. 1,373,108 granted March 29, 1921.

40. U.S. Patent No. 1,467,959 granted September 11, 1923; U.S. Patent No. 1,478,979 granted December 25, 1923; U.S. Patent No. 1,652,840 granted December 13, 1927.

41. Betsy H. Woodman, "The Salisbury Beach Dodgem: A Smashing Ride (1920–1980)," from The Essex Institute, Salem, MA. http://www.lusseautoscooters.com/html/legend_history.html. Accessed November 9, 2011.

42. U.S. Patent No. 1,013,792 granted January 2, 1912; U.S. Patent No. 1,112,305 granted September 29, 1914.

43. U.S. Patent No. 1,665,103 granted April 3, 1928.

44. Seth Gussow, "History of Bumper Cars," *Automobile Magazine* (November 1997), http://www.heart4theword.hubpages.com/hub/Bumper-Cars. Accessed November 9, 2011.

45. http:www.bumpercars.onice.com/. Accessed November 15, 2011.

46. "White City (amusement parks), http://en.wikipedia.org/wiki/White_City_(amusement_park).html. Accessed November 15, 2011.

47. "Amusement Parks," *Encyclopedia of Chicago*, http://encyclopedia.chicagohistory.org/pages/48.html. Accessed November 15, 2011.

48. David Nye, *The Technological Sublime* (Cambridge: MIT Press, 1994), p. 143.

49. "White City (Chicago)," http://en.wikipedia.org/wiki/White_City_(Chicago).html. Accessed November 15, 2011.

50. *Chicago Daily Tribune*, December 6, 1959.

51. "Amusement Parks," *Cleveland History*, http://ech.case.edu/ech-cgi/article.pl?id=AP1, Accessed January 29, 2009.

52. "Cedar Point," http://en.wikipedia.org/wiki/Cedar_Point. Accessed October 19, 2011.

53. "Cedar Point," http://www.cedarpoint.com/public/park/rides/coasters/index.cfm. Accessed October 19, 2011.

54. National Park Service, "The Glen Echo Amusement Park," http://www.nps.gov/nr/twhp/wwwlps/lessons/24glenecho/24facts2.html. Accessed October 8, 2011.

55. National Park Service, "Great Falls Park," http://www.nps.gov/grfa/historyculture/carousels.html. Accessed October 8, 2011.

56. National Park Service, "Pullen Park Carousel," http://www.nps.gov/nr/travel/raleigh/pul.html. Accessed October 8, 2011.

57. National Park Service, "Playland Amusement Park," http://tps.cr.nps.gov/nhl/detail.cfm?ResourceId=1835&ResourceType=District. Accessed October 10, 2011; "Playland" (New York), http://en.wikipedia.org/wik/Playland_(New_York). Accessed October 17, 2011.

58. "History of the Pier," http://brightonpier.co.uk/history-of-the-pier. Accessed October 17, 2011; Eric Abrahamson, "Amusement Parks & the California Coast," *California Historical Courier* (June–July 1986), 6.

59. "The Pike," http://en.wikipedia.org/wiki/The_Pike. Accessed October 19, 2011.

60. Abrahamson, "Amusement Parks & the California Coast," 6.

61. "Charles I .D. Looff," http://en.wikipedia.org/wiki/Charles_I_D_Looff. Accessed October 9, 2011.

62. National Park Service, "Santa Cruz Looff Carousel & Roller Coaster On the Beach Boardwalk," http://tps.cr.nps.gov/nhl/detail.cfm?ResourceId=1978&ResourceType=Structure. Accessed October 8, 2011.

63. "John D. Spreckels," http://en.wikipedia.org/wiki/John_D_Spreckels. Accessed October 16, 2011; "Belmont Park (San Diego)," http://en.wikipedia.org/wiki/Belmont_Park_(San_Diego). Accessed October 16, 2011; National Park Service, "Mission Beach Roller Coaster," http://tps.cr.nps.gov/nhl/detail.cfm?ResourceId=1746&ResourceType=Structure. Accessed October 10, 2011.

64. San Francisco Public Library, "Art at Zoo: Mr. Woodward's Gardens," http://sfpl.org/index.php?pg=2000119401. Accessed October 19, 2011.

65. San Francisco Public Library, "De Young's Delight: The Midwinter Fair," http://sfpl.org/index.php?pg=2000131901. Accessed October 19, 2011.

66. San Francisco Public Library, "Splish! Splash! At the Chutes," http://sfpl.org/.index.php?pg=2000131701. Accessed October 19, 2011.

67. San Francisco Public Library, "Chutes at the Beach," http://sfpl.org.index.php?pg=2000141301. Accessed October 19, 2011; National Park Service, "Archeological Stewardship Program at Lands End," http: www.nps.gov/goga/parknews/2011–0513.html. Accessed October 10, 2011.

68. San Francisco Public Library, "Chutes at the Beach"; "Charles I. D. Looff," http://en.wikipedia.org/wiki/Charles_I_D_Looff.

69. *San Francisco Chronicle*, July 3, 2005.

70. Ibid.

71. "Coney Island—Movie List," http://www.westland.net/coneyisland/articles/movielist.html. Accessed June 18, 2012.

72. Robert E. Snow and David E. Wright, "Coney Island: A Case Study in Popular Culture and Technical Change," *Journal of Popular Culture*, 9 (Spring 1976), 962.

73. Ibid., 964.

74. Ibid., 966–967.

75. "NYC Mayor: Coney Island 'is coming back, big time'," *San Francisco Chronicle*, February 16, 2010.

76. *Sydney Morning Herald*, April 2, 2010.

77. "IAAPA HISTORY: A History of IAAPA: The Evolution of Today's Premier Attractions Association and the Industry It Serves," http://www.iaapa.org/aboutus/history/. Accessed November 11, 2011.

78. "What's New at the Parks," *Sunset*, 190 (May 1993), 28; the following description relies heavily on Carroll Pursell, *Technology in Postwar America* (New York: Columbia University Press, 2007), pp. 111–113.

79. Margaret J. King, "Disneyland and Walt Disney World: Tradition al Values in Futuristic Form," *Journal of Popular Culture*, 15 (Summer 1981), 116.

80. Ibid., 121.

81. Sorkin, "See You in Disneyland," pp. 210, 216.

82. Tom Vanderbilt, "Mickey Goes to Town(s)," *Nation*, 261 (August 28/September 4, 1995), 198.

83. Quoted in Miller's obituary, *San Francisco Chronicle*, September 12, 2001. Ironically, he died just two weeks before a pair of jumbo jets were flown into the towers of the World Trade Center.

84. Sorkin, "See You in Disneyland," p. 228.

85. See his obituary in *New York Times*, December 14, 1993.

86. Quoted in King, "Disneyland and Walt Disney World," 121.

87. Vanderbilt, "Mickey Goes to Town(s)," 200.

88. Michael L. Smith, "Back to the Future: EPCOT, Camelot, and the History of Technology," *New Perspectives on Technology and American Culture*, ed. Bruce Sinclair (Philadelphia: American Philosophical Society, 1986), pp. 70–72. See also his "Making Time: Representations of Technology at the 1964 World's Fair," *The Power of Culture: Critical Essays in American History*, ed. Richard Wightman Fox and T. J. Jackson Lears (Chicago: University of Chicago Press, 1993), pp. 223–244.

89. Gary S. Cross and John K. Walton, *The Playful Crowd: Pleasure Places in the Twentieth Century* (New York: Columbia University Press, 2005), p. 5.

90. John Kasson, *Amusing the Millions: Coney Island at the Turn of the Century* (New York: Hill and Wang, 1978), pp. 6, 7.

Chapter Four: The Hobbyist

Epigraphs. "Model Building Builds Model Boys." Comet Industries Corp., *Meet Captain Comet "Jet Ace"* (n.p., n.d.), back cover. Quoted in Novella Carpenter, *Farm City: The Education of an Urban Farmer* (New York: Penguin Books, 2009), p. 41.

1. Steven M. Gelber, *Hobbies: Leisure and the Culture of Work in America* (New York: Columbia University Press, 1999), pp. 1, 2.

2. Ibid., p. 2.

3. Ibid., pp. 3, 5, 156.

4. Rachel P. Maines, *Hedonizing Technologies: Paths to Pleasure in Hobbies and Leisure* (Baltimore: Johns Hopkins University Press, 2009), pp. 7, 3.

5. Tine Kleif and Wendy Faulkner, "Boys and Their Toys: Men's Pleasures in Technology," in U. Pasero & A. Gottburgsen, eds., Wie natürlich ist Geschlecht? Gender und die Konstruktion von Natur und Technik [How Natural Are Sex and Gender? Gender and the Construction of Nature and Technology] (Wiesbaden: Westdeutscher Verlag, 2002), also available at docin.com/p-55359570.html.

6. The following relies upon Elizabeth Harris, *The Boy and His Press* (Washington, DC: Smithsonian Institution, 1992), passim.

7. Frank Lloyd Wright, *An Autobiography* (New York: Duell, Sloan and Pearce, 1943), pp. 33, 36.

8. Edward I. Pratt, "Boys as Aeroplane Modelers," *Illustrated World*, 20 (November 1913), 423–425.

9. Richard Butsch, "The Commodification of Leisure: The Case of the Model Airplane Hobby and Industry," *Qualitative Sociology*, 7 (Fall 1984), 217–235.

10. Aaron L. Alcorn, "Flying into Modernity: Model Airplanes, Consumer Culture, and the Making of Modern Boyhood in the Early Twentieth Century," *History and Technology*, 25 (June 2009), 116.

11. Ibid., 128.

12. Ibid., 136.

13. Butsch, "The Commodification of Leisure," 224–225.

14. Comet Industries Corp., *Meet Captain Comet "Jet Ace"* (n.p., n.d.), back cover.

15. Kristen Haring, *Ham Radio's Technical Culture* (Cambridge: MIT Press, 2007), pp. ix, xv; Susan J. Douglas, "Audio Outlaws: Radio and Phonograph Enthusiasts," *Possible Dreams: Enthusiasm for Technology in America*, ed. John L. Wright (Dearborn: Henry Ford Museum & Greenfield Village, 1992), pp. 44–59.

16. Douglas, "Audio Outlaws," passim.

17. Ibid., pp. 48–49.

18. Wayne Whipple and S. F. Aaron (Chicago: M.A. Donohue & Co., 1922), p. 7.

19. Douglas, "Audio Outlaws," p. 51.

20. Haring, *Ham Radio's Technical Culture*, p. ix; Kristen Haring, "The 'Free Men' of Ham Radio: How a Technical Hobby Provided Social and Spatial Distance," *Technology and Culture*, 44 (October 2003), 735–736.

21. Haring, "The "Free Men'," 740–741.

22. "Edward Bayard Heath," http://en.wikipedia.org/wiki/Edward_Bayard_Heath. Accessed February 28, 2013; Heath Kit Airplanes," http://www.hendersonkj.com/air planes/heath/index.html. Accessed November 29, 2011.

23. "Heathkit," http://en.wikipedia.org/wikie/Heathkit. Accessed November 29, 2011.

24. Douglas, "Audio Outlaws," pp. 52–53.

25. Cited in Keir Knightley, "'Turn it down!' She Shrieked: Gender, Domestic Space, and High Fidelity, 1948–59," *Popular Music*, 15, no. 2 (1996), 149–177.

26. "Heathkit," Wikipedia.

27. Lawrence M. Fisher, "Plug Is Pulled on Heathkits, Ending a Do-It-Yourself Era," *New York Times*, March 30, 1992.

28. Douglas, "Audio Outlaws," pp. 54–59.

29. John Jerome, *Truck: On Rebuilding a Worn Out Pickup, and Other Post-Technological Adventures* (Boston: Houghton Mifflin Co., 1977), p. 4.

30. J. T. Borhek, "Rods, Chopper, and Restorations: The Modification and Re-creation of Production Motor Vehicles in America," *Journal of Popular Culture*, 22 (Spring 1989), 101.

31. Quoted in James Sterngold, "Making the Jalopy an Ethnic Banner," *New York Times*, February 19, 2000.

32. "Hot Rods, Car Clubs and Drag Strips," *Information Report*, No. 74 (May 1955), Planning Advisory Service, American Society of Planning Officials, pp. 4–5.

33. Ibid.

34. H. F. Moorhouse, "The 'Work' Ethic and 'Leisure' Activity: The Hot Rod in Post-war America," *The Historical Meanings of Work*, ed. Patrick Joyce (Cambridge: Cambridge University Press, 1987), pp. 246–247, 251.

35. Tim Molloy, "Engineers Modify Hybrid Cars to Get Up to 250 mpg," *San Francisco Chronicle*, August 13, 2005.

36. Gelber, *Hobbies*, pp. 185, 207.

37. Frederick Winslow Taylor, *Principles of Scientific Management* (New York: Harper and Brothers, 1911), pp. 43–44.

38. Gelber, *Hobbies*, p. 205.

39. Carroll Pursell, *Technology in Postwar America* (New York: Columbia University Press, 2007), p. 22.

40. Elaine Tyler May, *Homeward Bound: American Families in the Cold War Era* (New York: Basic Books, 1988), p. 165.

41. "Cinema History from the Cold War: Walt Builds a Family Fallout Shelter," http://www.atomictheater.com/waltfalloutshelter.htm. Accessed April 24, 2014.

42. U.S. Patent No. 1,245,860 granted November 6, 1917; "History of Black & Decker," www.blackandecker.co.uk/about/highlights. Accessed August 16, 2005.

43. Carolyn M. Goldstein, *Do It Yourself: Home Improvement in 20th-Century America* (New York: Princeton Architectural Press, 1998), pp. 49–50.

44. U.S. Patent No. 1,905,462 granted April 25, 1933.

45. "Husky (tools)," http://ed.wikipedia.org/wiki/Husky_(tools). Accessed June 3, 2013.

46. Quoted in Peter H. Lewis, "Smart Tools for the Home with Chips, Lasers, L.C.D.'s," *New York Times*, April 20, 2000.

47. Goldstein, *Do It Yourself*, pp. 51–52.

48. Ibid., pp. 54–55.

49. Ibid., p. 64.

50. "The Home Depot," http://en.wikipedia.org/wiki/The_Home_Depot.

51. Nancy Davis Kho, "TechShop SF Open House for DIYers," *San Francisco Chronicle*, February 16, 2011.

52. Maines, *Hedonizing Technologies*, p. 6.

53. Pilar Guzman, "Hey, Man, What's for Dinner," *New York Times*, August 28, 2002.

54. *Guardian*, April 10, 2013.

55. Denise Wasley, "If You Can't Stand the Housework, Don't Hide in the Kitchen," *Sydney Morning Herald*, February 17, 2005.

56. Carpenter, *Farm City*, p. 41.

Chapter Five: Games and Sports

Epigraphs. U.S. Patent No. 335,656 granted February 9, 1886. *Sydney Morning Herald*, May 14–15, 2005.

1. Quoted in David Nasaw, *Children of the City: At Work & At Play* (New York: Anchor Press/Doubleday, 1985), p. 30.

2. Quoted in John Rickards Betts, "The Technological Revolution and the Rise of Sport, 1850–1900," *Mississippi Valley Historical Review*, 40 (September 1953), 235.

3. Ibid., 237–238.

4. "Albert Spalding," http://en-wikipedia.org/wiki/Albert_Spalding. Accessed January 29, 2012.

5. Betts, "Technological Revolution," 239–240.

6. Ibid., 242.

7. Stephen Hardy, "'Adopted by All the Leading Clubs': Sporting Goods and the Shaping of Leisure, 1800–1900," in *For Fun and Profit: The Transformation of Leisure into Consumption*, ed. Richard Butsch (Philadelphia: Temple University Press, 1990), p. 75; Betts, "Technological Revolution," 244.

8. Hardy, "Adopted by All the Leading Clubs," pp. 74, 78.

9. Ibid., p. 80; "Albert Spalding," http://en.wikipedia.org/wiki/Albert_Spalding. Accessed January 29, 2012.

10. Hardy, "Adopted by All the Leading Clubs," pp. 81–82.

11. Ibid., pp. 82–83.

12. Ibid., pp. 85, 87, 88.

13. Ibid., pp. 91, 92, 93.

14. Rick Hinden, "Take Me Back to the Ball Park," *Historic Preservation*, 31 (July–August 1979), 47.

15. Ibid., 44, 47, 49–50; "Reliant Astrodome," en.wikipedia.org/wiki/Reliant_Astrodome. Accessed November 6, 2013.

16. "Luther Halsey Gulick," http://www.infed.org/thinkers/gulick.htm. Accessed October 17, 2005.

17. *New York Times*, November 12, 1890.

18. "James Naismith," http://kshs.org/portraits/naismith_james.htm. Accessed July 23, 2010; "Basketball 'Bible' Auction Sets Sports Memorabilia Record," CNN.com, edition.cnn.com/2010/SPORT/12/10/basketball.rules.auction/. Accessed December 9, 2013.

19. *New York Times*, December 6, 2006.

20. W J McGee, "Fifty Years of American Science," *Atlantic Monthly*, 82 (September 1898), 311–312.

21. Andrew Ritchie, *Major Taylor: The Extraordinary Career of a Champion Bicycle Racer* (Baltimore: Johns Hopkins University Press, 1996).

22. John Markoff, "Aerospace on 2 Wheels: Lightweight and Strong," *New York Times*, October 20, 2003.

23. "History of Tennis," hppt://en.wikipedia.org/wiki/History_of_tennis. Accessed August 22, 2013.

24. U.S. Patent No. 335,656 dated February 9, 1886.

25. U.S. Patent No. 401,082 dated April 9, 1889.

26. "About.com Tennis," http://tennis.about.com/od/racquetsballsstringing/a/evolmodracquet.html. Accessed September 3, 2013.

27. "Tennis ball," http://en.wikipedia.org/wiki/Tennis_ball. Accessed August 22, 2013.

28. "History of Golf," http://en.wikipedia.org/wiki/History_of_golf. Accessed September 26, 2013.

29. "The History of the Golf Club," hhtp://home.aone.net.au/Byzantium/golf/ghistory.html. Accessed September 8, 2005; U.S. Patent No. 622,834 issued April 11, 1899.

30. Quoted in Frank Barkley Copley, *Frederick W. Taylor: Father of Scientific Management* (New York: Harper and Brothers, 1923), II, pp. 205–206.

31. U.S. Patent No.732,136 and No. 732,137, both dated June 30, 1903; Copley, *Frederick Winslow Taylor*, II, pp. 216–219; "Broomhandle Putters Set to Be Banned," ABC News, http//www.abc.net.au/news/2012–11–29/broomhandle-putters-set-to-be-banned/4398024. Accessed November 29, 2012.

32. "The History of the Golf Club."

33. "Hootie Defends Augusta's Changes," sports.espn.go.com/golf/news/story?id= 2398253. Accessed October 8, 2013.

34. Greg Duke, "Golf Gadgets to Improve Your Game-CNN.com," http.//edition.cnn.com/2012/02/29/sport/golf/golf-products-2012/index.html. Accessed March 1, 2012.

35. Marc Saltzman, "Review: Tee Off with Two Solid Gold Offerings," http://www.cnn.com/2007/TECH/fun.games/04/06/golf.games/index.html. Accessed April 13, 2007; Daniel Falton, "Club Class," *The (Sydney) Magazine* (April 2006), 99.

36. John Margolies et al., *John Margolies's Miniature Golf* (New York: Abbeville Press, 1987), p. 14.

37. U.S. Patent No. 1,559,520 granted October 27, 1925.

38. Margolies, *Miniature Golf*, pp. 20–21.

39. Ibid., pp. 22–23; U.S. Patents Des. 83,497; 83,498; 83,501; 83,505; 83,506 all granted on March 3, 1931.

40. Margolies, *Miniature Golf*, pp. 23–24.

41. Quoted in ibid., pp. 74–75.

42. Obituaries of Ralph J. Lomma in *New York Times*, September 17, 2011 and *Washington Post*, September 16, 2011.

43. Margolies, *Miniature Golf*, p. 89.

44. "The History of the Athletic Shoe," http://www.runtheplanet.com/resources/historical/athleticshoes.asp. Accessed March 23, 2011.

45. "German Language," http://german.about.com/library/blerf_dassler.html. Accessed March 23, 2011.

46. *Sydney Sun-Herald*, Sunday Life section, January 6, 2008.

47. "Best Foot Forward," *Sydney Morning Herald*, Good Weekend section, October 1, 2011.

48. *San Francisco Chronicle*, August 16, 2008.

49. *Sydney Morning Herald*, May 4–5, 2005.

50. *San Francisco Chronicle*, May 24, 2006; *Sydney Morning Herald*, February 26, 2007.

51. *Sydney Morning Herald*, March 12, 2009.

52. "Best Foot Forward."

53. "Tights May Give Runners a Lift," *Sydney Morning Herald*, June 14, 2006.

54. "Olympic Athletes Go High Tech," *Sydney Morning Herald*, April 14, 2008.

55. ABC News (Australia), http://www.abc.net.au/news/2012–05–09/british-woman-makes-marathon-history-in-bionic suit/3999710. Accessed May 9, 2012.

56. *Sydney Morning Herald*, March 22, 2007, www.speedo.com/speedo_brand/insidespeedo/history/index.html. Accessed December 6, 2013.

57. "LZR Racer," http://en.wikipedia.org/wiki/LRZ_Racer. Accessed December 7, 2013.

58. Meredith May, "Keeping a Closer Eye on Athletes. New Devices Help Track Winners, Losers at Games," *San Francisco Chronicle*, August 23, 2004.

59. Ibid.

60. "Goal-line Technology Enters Final Testing," March 3, 2012, http://edition.cnn.com/2012/03/03/sports/football/goal-line-tech-fifa/index.html. Accessed March 4, 2012; "FIFA confirms Goal-line Technology at World Cup," http://www.abc.net.au/news/2013–02–19/fifa-confirms-goal-line-technology. Accessed February 20, 2013.

61. "Umpire Review System," http://en.wikipedia.org/wiki/Umpire_Decision_Review_System. Accessed September 13, 2013; "Ashes 2013: ICC Reportedly Investigating Use of Silicon Tape on Bat Edges to Evade Hot Spot," http://www.abc.net.au/news/2013–08–07/icc-reportedly-probing-hot-spot-evasion-tactics/4872072. Accessed August 7, 2013.

62. *Sydney Morning Herald*, October 10, 2013.

63. "Hawk-Eye," http://en.wikipedia.org/wiki/Hawk-Eye. Accessed January 10, 2008.

64. Ibid.

65. Benny Evangelista, "LiveLine Technology to Keep Tabs on America's Cup," *SFGate*, January 1, 2012.

Chapter Six: Extreme and (Sometimes) Impolite Sports

Epigraphs. *Extreme Landscapes of Leisure: Not a Hap-Hazardous Sport* (Farnham: Ashgate Publishing, 2011), p. 73. "IAAPA HISTORY," http://www.iaapa.org/aboutus/history/. Accessed November 11, 2011.

1. "Extreme sport," Wikipedia.

2. "X Games History," http://www.tqnyc.org/NYC030417/html/xgameshistory.html.

3. "Tahoe Vies for X Games," SFGate.com blog, posted January 17, 2013.

4. "X games," http://en.wikipedia.org/wiki/X_Games. Accessed January 22, 2013.

5. USPS Postal News, June 25, 1999, http:..www.espneventmedia.com/uploads/application/XGamesStamps.pdf. Accessed February 1, 2012.

6. Laviolette, *Extreme Landscapes of Leisure*, pp. 46–47.

7. "The Surfboard," July 22, 2002, http://www.npr.org/programs/morning/features/patc/surfboard/.

8. Ben Finney and James D. Houston, *Surfing: A History of the Ancient Hawaiian Sport*, rev. ed. (San Francisco: Pomegranate Artbooks, 1996), p. 81.

9. Ibid.

10. Jack London, *The Cruise of the Snark* (1911), found at http://www.jacklondon.net/surfing3.html.

11. Finney and Houston, *Surfing*, p. 81; "Duke Kahanamoku," http://www.wikipedia.org/wiki/Duke_Kahanamoku.

12. Burl Burlingame, "Original 'Gidget,' teen icon of the '60s, turns 60," *Honolulu Star-Bulletin*, January 22, 2001.

13. Finney and Houston, *Surfing*, p. 83.

14. Ibid., p. 7.

15. "History of the Surfboard," http://360guide.info/surfing/history-of-the-surfboard.html.

16. Alex McDonald and Deborah Smith, "Hey, Dude, Who Stole the Waves," *Sydney Morning Herald*, February 7, 2009.

17. "History of the Surfboard."

18. "Bruce 'Snake' Gabrielson Shaper for Wave Trek Surfboards—1970s," http://blackmagic.com/ses/book/his-c.html.

19. "Hobie Alter. He Started Out Shaping Surfboards, He Ended Up Shaping a Culture," http://www.hobie.com/history/.

20. Mike Anton, "Wave Goodbye to Handmade Surfboards," *San Francisco Chronicle*, July 15, 2007.

21. Ibid.

22. Ibid.

23. (Sydney) *Sun-Herald*, February 4, 2007.

24. Michael Taylor, "Hugh Bradner, UC's Inventor of Wetsuit, Dies," *San Francisco Chronicle*, May 11, 2008; "A Waterman's Tale: The True Inventor of the Wetsuit (Part 2)," http://www.surfpulse.com/2007/features/i_a_waterman_s_tale_090.

25. "History of the Wetsuit," January 15, 2007, http://360guide.info/wetsuits/wetsuit-history.html.

26. "The Story Behind Body Glove," http://www.divinghistory.com/historyofbodyglove.html.

27. Finney and Houston, *Surfing*, p. 7.

28. *New York Times*, May 13, 1909.

29. Ed Machado, "Waimea Bay Birthplace of a New Type of Man-made Wave," http://www.surfervillage.com/print.asp?Id_news=15598.html.

30. Jim Benning, "Even for Veteran Surfers, an Artificial Wave can Bruise Bones and Egoes. And now comes Bruticus Maximus," *Los Angeles Times*, May 27, 2005.

31. "Take Five: Wave Pools," (Sydney) *Sun-Herald*, January 29, 2006.

32. Benning, "Even for Veteran Surfers, an Artificial Wave Can Bruise Bones and Egoes."

33. *Sun-Herald* (Sydney), January 29, 2006.

34. "Computer Device Makes Big Waves," http://www.cnn.com/2005/TECH/07/05/surfing.reefs.

35. Cameron Lawrence, "How Skateboarding Works," http://entertainment.howstuff works.com/skateboarding.html.

36. Ibid.

37. G. Beato, "The Lords of Dogtown," *Spin*, March 1999, http://www.angelfire.com/ca/alva3/spin.html.

38. Ibid.

39. Catherine Ellsworth, "US Skateboarders Pool Their Talents," *The Age*, January 5, 2009.

40. "The History of Vert Ramps," http://www.angelfire.com/mn/skate68/history ofvert.html.

41. Lawrence, "How Skateboarding Works."

42. From the film *Dogtown and Z-Boys*, 2002.

43. Iain Borden, *Skateboarding, Space and the City: Architecture and the Body* (Oxford: Berg, 2001), p. 2.

44. "Sports History: Skateboarding," http://www.hickoksports.com/history/skate boarding.shtml.

45. "Skateboard History," http://www.tqnyc.org/NYC030417/html/skateboardhistory .html.

46. Mark Simborg, "Shredding Again: In Which We Chronicle Midlife Skate-boarders and Their Return, to Be Schooled by 6-year-olds," *San Francisco Chronicle*, October 21, 2007.

47. Miho Hosaka, "The Story of Dreamland Skateparks," http://www.switchmagazine .com/skateboard_storys/dreamland_skateparks.html.

48. "The Dreamland 'Dreamteam'," http://www.dreamlandskaeparks.com/employees/aboutus.html.

49. Matt Higgins, "A Skateboarding Ramp Reaches for the Sky," *New York Times*, November 1, 2006.

50. *Sydney Morning Herald*, February 9, 2009.

51. "It's Sport, It's Extreme, It's . . . Ironing," BBC News, August 7, 2002, http://news .bbc.co.uk.

52. Finney and Houston, *Surfing*, p. 85.

53. Laura Zelasnic, "S. Newman Darby Windsurfing Collection, 1944–1998," Smithsonian Archives, August 1999, http://amhistory.si.edu/archives/. Accessed January 29, 2013.

54. Interview with Drake in *American Windsurfer Magazine*, 4, no. 4 (1996).

55. http://www.windsurfing-academy.com/information_bank/history. Accessed February 8, 2008.

56. Quoted in Steve Rubenstein, "Kiteboarders in S.F., Hoping Sport Will Woar," *San Francisco Chronicle*, June 11, 2008.

57. Ibid.

58. http://www.tqnyc.org/NYC030417/html/snowboardhistory.html. Accessed May 26, 2005.

59. U.S. Patent No. 3,378,274 issued April 16, 1968; "Inventor Sherman Poppen, Snurfing His Way into History, http://blog.americanhistory.si.edu/osaycanyousee/2009/09/inventor-sherman-popper-snurfing-his-way-into-history.html. Accessed March 13, 2013.

60. Paul J. MacArthur, "The Top Ten Important Moments in Snowboarding History," Smithsonian.com, February 5, 2010; U.S. Patent No. 3,782,744 issued January 1, 1974.

61. MacArthur, "Top Ten"; "History of Snowboarding," http://en.wikipedia.org/wiki/Snowboarding accessed January 31, 2013.

62. MacArthur, "Top Ten."

63. "Hawaiians Slide into Dangerous Sport," CNN.com, August 21, 2005.

64. "Lilienthal Glider," http://www.claiforniasciencecenter/org/Exhibits/AirAndSpace/AirAndAircraft/LilienthalGlider/LilienthalGlider.php.

65. Carl Bates, "How to Build a Glider," *The Boy Mechanic: 700 Things for Boys to Do* (Chicago: Popular Mechanics Press, 1913).

66. John Underwood, "Volmer Jensen—Remembered," Soaring Society of America, http://www.ssa.org/myhome.asp?id=662&mbr=5811273455; "Volmer VJ-11 Hang Glide 'Information Package'," http://esoaring.com/volmer_vj-11.htm.

67. U.S. Patent No. 2,546,078 awarded March 20, 1951; U.S. Patent No. 2,751,172 awarded June 19, 1956.

68. Douglas Martin, "F. M. Rogallo, Father of Hang Gliding, Dies at 97," *New York Times*, September 5, 2009.

69. Ibid.; obituary of Bill Bennett, *Sydney Morning Herald*, November 26, 2004.

70. "Delta Wing Phoenix VI B Jr., Collections Search Center, Smithsonian Institution, http://collections.si.edu/search/record/nasm_A19840715000?print-yes. Accessed May 10, 2013; obituary of Bennett in *New York Times*, November 17, 2004.

71. Doug George, "Fly Boys," Lifestyles section, *Chicago Tribune*, August 27, 2004.

72. Ibid.

73. Cameron Lawrence, "How Parkour Works," http://adventure.howstuffworks.com/outdoor-activities/urban-sports/. Accessed August 1, 2012.

74. Dan Edwardes, "Parkour History," http://www.parkourgenerations.com/article/parkour-history. Accessed August 1, 2012.

75. Erwan LeCorre, quoted in http:///e..wikipedia.org/wiki/Parkopur. Accessed August 1, 2012.

76. Quoted in "Extreme Sport," http://en.wikipedia.org/wiki/Extreme_sport. Accessed January 31, 2012.

77. Wolfgang Schivelbusch, *The Railway Journey: The Industrialization of Time and Space in the 19th Century* (Berkeley: University of California Press, 1986), pp. 53, 54.

78. J. B. Jackson, "The Abstract World of the Hot-Rodder," *Changing Rural Landscapes*, ed. Ervin H. Zube and Margaret J. Zube (Amherst: University of Massachusetts Press, 1977), pp. 146–147.

Chapter Seven: Electronic Games

Epigraph. Mary Bellis, "Spacewar Online," http://inventors.about.com/library/weekly/aa090198.htm. Accessed January 3, 2007.

1. "Bagatelle," en.wikipedia.org/wiki/Bagatelle. Accessed December 10, 2013; www.loc.gov/pictures/ten/2008661662.

2. U.S. Patent No. 115,357 granted May 30, 1871; U.S. Patent No. 603,738 patented May 10, 1898.

3. "Pinball," http://en.wikipedia.org/wiki/Pinball. Accessed December 14, 2013.

4. Ibid.

5. Quoted in Frederic D. Schwartz, "The Patriarch of Pong," *American Heritage Invention and Technology*, 6 (Fall 1990), 64.

6. John Markoff, "A Long Time Ago, in a Lab Far Away . . .," *New York Times*, January 28, 2002.

7. Quoted in ibid.

8. Aphra Kerr, *The Business and Culture of Digital Games: Gamework/Gameplay* (London: Sage Publications, 2006), p. 14; Mary Bellis, "Spacewar: The First Computer Game Invented by Steve Russell," http://inventors.about.com/library/weekly/aa090198.htm. Accessed March 1, 2007.

9. Thomas P. Hughes, *American Genesis: A Century of Invention and Technological Enthusiasm, 1870–1970* (New York: Viking, 1989), p. 24.

10. Quoted in Markoff, "A Long Time Ago."

11. Ibid.; Steven Poole, *Trigger Happy: Videogames and the Entertainment Revolution* (New York: Arcade Publishing, 2000), p. 19.

12. "Welcome to PONG-Story," http://www.pong-story.com/intro.htm; Ralph H. Baer, "Foreword," to Mark J. P. Wolf, *The Medium of the Video Game* (Austin: University of Texas Press, 2001), pp. xii–xiii.

13. Poole, *Trigger Happy*, p. 20.

14. "Fun and Games," *Fortune*, 107 (May 2, 1983), 164.

15. Steven Levy, *Hackers: Heroes of the Computer Revolution* (New York: Anchor Press/Doubleday, 1984), p. 325.

16. Ibid., pp. 324–325, 352–353.

17. "Fun and Games," 164.

18. Jeffrey B. Mahan, "Federal Copyright Law in the Computer Era: Protection for the Authors of Video Games," *University of Puget Sound Law Review*, 425 (1983–1984), 430.

19. Ibid., 428, 433–434.

20. Interview with John Harris at http://www.dadgum.com/halcyon/BOOK/HARRIS. HTML.

21. "Sega," http://en.wikipedia.org/wiki/Sega. Accessed December 19, 2013.

22. "Taito," http://en.wikipedia.org/wiki/Taito_Corporation. Accessed December 19, 2013.

23. "Namco," http://en.wikipedia.org/wiki/Namco. Accessed December 19, 2013. *Sunday Life* [Sydney] *Sun-Herald Magazine*, January 14, 2007), 10.

24. "Pac-Man Chomps Up Milestone," http://www.cnn.com/2005/TECH/fun.games/06/15/pac.man.25.ap/index.html.

25. Quoted in Keith Stuart, "Farewell, Mr. Mario—former Nintendo Head Hiroshi Yamauchi Dies," *Guardian*, September 20, 2013.

26. Ibid.

27. "Hiroshi Yamauchi: The Man Who Brought You Nintendo," *Sydney Morning Herald*, September 20, 2013.

28. "Nintendo," http://en.wikipedia.org/wiki/Nintendo. Accessed December 19, 2013.

29. Marshall A. Fey, "Charles Fey and San Francisco's Liberty Bell Slot Machine," *California Historical Quarterly*, 54 (Spring 1975), 57; "Charles Fey," http://www.casinoman.net/content/slot-machines/charles-fey.asp. Accessed December 12, 2013.

30. Ellen Paris, "Call and Raise," *Forbes*, 130 (August 30, 1982), 50–51.

31. Bart Eisenberg, "The New 'One-Arm Bandits': Today's Slot Machines Are Built Like PCs, Programmed Like Video Games," Pacific Connection, February, 2004 issue of *Software Design*, http:www.gihyo.co.jp/magazine/SD/pacific/SD_0402.html.

32. The following discussion is taken from her article, "Digital Gambling: The Coincidence of Desire and Design," *Annals of the AAPSS*, 597 (January 2005), 65–81.

33. Ibid., 67, 69.

34. Ibid., 72, 73, 77.

35. Ibid., 78–79.

36. Thomas Crampton, "For France, Video Games Are as Artful as Cinema," *New York Times*, November 6, 2006.

37. Quoted in Duane Brown, "The Art & Science of Public Relations—It Doesn't Have to Be Russian Roulette If You Don't Want It to Be," *Gamasutra*, March 15, 2007.

38. Alxa Moses and Elicia Murray, "Good Game But Is It Art? *Sydney Morning Herald*, September 3, 2006.

39. Aaron Smuts, "Video Games and the Philosophy of Art," *Aesthetics*, http://www.aesthetics-online.org/ideas/smuts.html.

40. *Sydney Morning Herald*, June 7, 2007.

41. Quoted in ibid.

42. Quoted in *Sydney Morning Herald*, June 25, 2007.

43. *Sydney Morning Herald*, June 14, 2007.

44. Quoted in *Sydney Morning Herald*, Metro Section, June 15–21, 2007 and June 25, 2007.

45. http://www.stanford.edu/class/sts145/; Stephan Schmitt, "Half a Century of Digital Gaming: *Game On*, at the Science Museum, London, 21 October 2006–25 February 2007," *Technology and Culture*, 48 (July 2007), 582–588; "How to Become a Video Games Designer, and What It Takes to Do Well in Video Game Design," http://www.adigitaldreamer.com/articles/becomeavideogamedesigner.html; "Game Art & Design," http://www.aionline.edu/degrees/game-art-design.html; "Computer Games Design and Technology," http://www.cms.livjm.ac.uk/gdtw/GDT2005.html; "Warwick Video Game Design Society," http://www.warwickgamedesign.co.uk/frontend.php?fepage+Home.

46. "Governor Signs Bills Involving Video Games, Sports Supplements," *San Francisco Chronicle*, October 7, 2005.

47. Julia Baird, "Thrill of the Kill: The Other Tragedy in Iraq," *Sydney Morning Herald*, November 17, 2005.

48. Jon Boone, "Taliban Retaliate After Prince Harry Compares Fighting to a Video Game," *Guardian*, January 22, 2013.

49. "Shoot to Thrill," *Sydney Morning Herald*, November 22, 2007.

50. Quoted in "Video Game 'Mods' Exploited by Islamists," *Sydney Morning Herald*, May 5, 2006.

51. "Leak: Government Spies Snooped in 'Warcraft,' Other Games," http://edition. cnn.com/2013/12/09/tech/web/nas-spying-video-games/index.htms?htp+hp_t3. Accessed December 10, 2013.

52. "Programmers At Work: Toru Iwatani, 1986 PacMan Designer," http://program mersatwork.wordpress.com/toru-iwatani-1986-packman-designer/. Accessed December 19, 2013. See also Nina B. Huntemann and Matthew Thomas Payne, eds., *Joystick Soldiers: The Politics of Play in Military Video Games* (New York: Routledge, 2009).

53. "The Electronic Gender Gap," *Washington Post*, November 29, 1984; "It's a Man's (virtual) World," *Sydney Morning Herald*, July 2, 2012.

54. "It's a Man's (virtual) World."

55. "Top 9 Greatest Video Game Heroines," *San Francisco Chronicle*, February 17, 2009; "Girl Power," *Sydney Morning Herald*, April 20, 2006.

56. Justine Cassell and Henry Jenkins, "Chess for Girls? Feminism and Computer Games," *From Barbie to Mortal Kombat: Gender and Computer Games*, ed. Justine Cassell and Henry Jenkins (Cambridge: MIT Press, 1998), p. 4.

57. *New York Times*, February 17, 1997.

58. "Barbie Boots Up," *Time*, November 11, 1996.

59. "Drawing Minorities into Gaming," CNN.com, August 5, 2005.

Suggestions for Further Reading

Playing with Technology

Chudacoff, Howard P. *Children at Play: An American History*. New York: New York University Press, 2007.

Cross, Gary. "Play in America from Pilgrims and Patriots to Kid Jocks and Joystick Jockeys: Or How Play Mirrors Social Change." *American Journal of Play* (Summer 2008), 8–47.

Horowitz, Roger, ed. *Boys and Their Toys? Masculinity, Class, and Technology in America*. New York: Routledge, 2001.

Huizinga, J. *Homo Ludens: A Study of the Play-Element in Culture*. London: Routledge & Kegan Paul, 1949.

Konner, Melvin. *The Evolution of Childhood: Relationships, Emotion, Mind*. Cambridge: Harvard University Press, 2010.

Mintz, Steven. "Play and the History of American Childhood: An Interview with Steven Mintz." *American Journal of Play* (Fall 2010), 143–156.

Toys for Girls and Boys

Andress, Carol. "War Toys and the Peace Movement." *Journal of Social Issues*, 25 (January 1969), 85–100.

Attfield, Judy. "Barbie and Action Man: Adult Toys for Girls and Boys, 1959–93." In *The Gendered Object*, ed. Pat Kirkham (Manchester: Manchester University Press, 1996), pp. 80–89.

Ball, Donald W. "Toward a Sociology of Toys: Inanimate Objects, Socialization, and the Demography of the Doll World." *Sociological Quarterly*, 8 (Autumn 1967), pp. 447–458.

Bergen, Doris. "Play as the Learning Medium for Future Scientists, Mathematicians, and Engineers." *American Journal of Play* (Spring 2009), 413–428.

Cho, Erin K. "Lincoln Logs" Toying with the Frontier Myth." *History Today*, 43 (April 1993), 31–34.

Cross, Gary. *Kids' Stuff: Toys and the Changing World of American Childhood*. Cambridge: Harvard University Press, 1997.

Dodgson, Mark, David Gann, and Ammon Salter. *Think, Play, Do: Technology, Innovation, and Organization*. Oxford: Oxford University Press, 2005.

Ferguson, Eugene S. *Engineering and the Mind's Eye*. Cambridge: MIT Press, 1992.

Fleming, Dan. *Powerplay: Toys as Popular Culture*. Manchester: Manchester University Press, 1996.

Formanek-Brunell, Miriam. *Made to Play House: Dolls and the Commercialization of American Girlhood, 1830–1930*. New Haven: Yale University Press, 1993.

Gilbert, A. C. and Marshall McClintock. *The Man Who Lives in Paradise*. New York: Heimburger House, 1954.

Kline, Stephen. *Out of the Garden: Toys, TV, and Children's Culture in the Age of Marketing*. London: Verso, 1993.

Lauwaert, Maaike. "Playing Outside the Box—On LEGO Toys and the Changing World of Construction Play." *History and Technology*, 24 (September 2008), 221–237.

McReavy, Anthony. *The Toy Story: The Life and Times of Frank Hornby*. London: Ebury Press, 2002.

Petroski, Henry. "Work and Play." *American Scientist*, 87 (May–June 1999), 208–212.

Scott, Sharon M. *Toys and American Culture: An Encyclopedia*. Santa Barbara: Greenwood, 2009.

Varney, Wendy. "Of Men and Machines: Images of Masculinity in Boys' Toys." *Feminist Studies*, 28 (Spring 2002), 153–174.

The Safe and Rational Playground

American, Sadie. "The Movement for Small Playgrounds." *American Journal of Sociology*, 4 (1898), 159–170.

Cohen, Ronald D. "Child-Saving and Progressivism, 1885–1915." In *American Childhood: A Research Guide and Historical Handbook*, ed. Joseph M. Hawes and N. Ray Hiner (Westport: Greenwood Press, 1985), pp. 273–309.

Dattner, Richard. *Design for Play*. Cambridge: MIT Press, 1974.

Gil, Tim. *No Fear: Growing Up in a Risk Adverse Society*. London: Calpuste Gulbenkian Foundation, 2007.

Goodman, Cary. *Choosing Sides: Playground and Street Life on the Lower East Side*. New York: Schocken Books, 1979.

Guttmann, Allen. "The Progressive Era Appropriations of Children's Play." *Journal of the History of Childhood and Youth*, 3 (Spring 2010), 147–151.

Hayward, D. Geoffrey, Marilyn Rothenberger, and Robert R. Beasley. "Children's Play and Urban Playground Environments: A Comparison of Traditional, Contemporary, and Adventure Playground Types." *Environment and Behavior*, 6 (June 1974), 131–168.

Kozlovsky, Roy. "Adventure Playgrounds and Postwar Reconstruction." In *Designing Modern Childhood. History, Space, and the Material Culture of Children: An International Reader*, ed. Marta Gutman and Ning de Conick-Smith (New Brunswick: Rutgers University Press, 2007), pp. 171–206.

Lady Allen of Hurtwood. *Planning for Play*. London: Thames and Hudson, 1968.

Meehan, Thomas. "Creative (and mostly upper-middle-class) Playthings." *Saturday Review*, 55 (December 16, 1972), 42–47.

Mitchell, William Brian. "We Want a Play Factory:" Scientific Management and Urban Playgrounds, 1907–1929: The Example of Cleveland, Ohio. Master's thesis, History, Case Western Reserve University, May 2000.

Opie, Peter and Iona Opie. *Children's Games in Street and Playground*. Oxford: Oxford University Press, 1970.

Rosenzweig, Roy. *Eight Hours for What We Will: Workers and Leisure in an Industrial City, 1870–1920*. Cambridge: Cambridge University Press, 1983.

Rosin, Hanna. "The Overprotected Kid." *Atlantic* (March 2014), 74–84.

Sammond, Nicholas. *Babes in Tomorrowland: Walt Disney and the Making of the American Child*. Durham: Duke University Press, 2005.

Solomon, Susan G. *American Playgrounds: Revitalizing Community Space*. Hanover: University Press of New England, 2005.

From Pleasure Gardens to Fun Factories

Anderson, Norman D. and Walter R. Brown. *Ferris Wheels*. New York: Pantheon Books, 1983.

Braithwaite, David. *Fairground Architecture*. London: Hugh Evelyn, 1968.

Cross, Gary S. and John K. Walton. *The Playful Crowd: Pleasure Places in the Twentieth Century*. New York: Columbia University Press, 2005.

Fried, Frederick. "Last Ride for the Carousel Figures?" *Historic Preservation*, 29 (July–September 1977), 22–27.

Kasson, John. *Amusing the Millions: Coney Island at the Turn of the Century*. New York: Hill and Wang, 1978.

King, Margaret J. "Disneyland and Walt Disney World: Traditional Values in Futuristic Form." *Journal of Popular Culture*, 15 (Summer 1981), 116–140.

Mohun, Arwen P. "Designed for Thrills and Safety: Amusement Parks and the Commodification of Risk, 1880–1929." *Journal of Design History*, 14, no. 4 (2001), 291–306.

Mohun, Arwen P. "Amusement Parks for the World: The Export of American Technology and Know-How, 1900–1939." *Icon*, 19 (2013), 100–112.

Peiss, Kathy. *Cheap Amusements: Working Woman and Leisure in Turn-of-the-Century New York*. Philadelphia: Temple University Press, 1986.

Searles, William H. "The Ferris Wheel." *Journal of the Association of Engineering Societies*, 12 (November 1893), 614–623.

Smith, Michael L. "Back to the Future: EPCOT, Camelot, and the History of Technology." *New Perspectives on Technology and American Culture*, ed. Bruce Sinclair (Philadelphia: American Philosophical Society, 1986), pp. 69–79.

Snow, Robert E. and David E. Wright. "Coney Island: A Case Study in Popular Culture and Technical Change." *Journal of Popular Culture*, 9 (Spring 1976), 960–975.

Sorkin, Michael. "See You in Disneyland." *Variations on a Theme Park: The New American City and the End of Public Space*, ed. Michael Sorkin (New York: Noonday Press, 1992), pp. 205–232.

Vanderbilt, Tom. "Mickey Goes To Town(s)." *Nation*, 261 (August 28/September 4, 1995), 197–200.

Willey, Day Allen. "The Trolley-Park." *Cosmopolitan*, 33 (July 1902), pp. 265–272.

The Hobbyist

Alcorn, Aaron. "Flying into Modernity: Model Airplanes, Consumer Culture, and the Making of Modern Boyhood in The Early Twentieth Century." *History and Technology*, 25 (June 2009), 115–146.

Borhek, J. T. "Rods, Chopper, and Restorations: The Modification and Re-creation of Production Motor Vehicles in America." *Journal of Popular Culture*, 22 (Spring 1989), 97–107.

Butsch, Richard. "The Commodification of Leisure: The Case of the Model Airplane Hobby and Industry." *Qualitative Sociology* 7 (Fall 1984), 217–235.

Douglas, Susan J. "Audio Outlaws: Radio and Phonograph Enthusiasts." In *Possible Dreams: Enthusiasm for Technology in America*, ed. John L. Wright (Dearborn: Henry Ford Museum & Greenfield Village, 1992), pp. 44–59.

Gelber, Steven M. "Do-It-Yourself: Constructing, Repairing and Maintaining Domestic Masculinity." *American Quarterly*, 49 (March 1997), 66–112.

Gelber, Steven M. *Hobbies: Leisure and the Culture of Work in America*. New York: Columbia University Press, 1999.

Goldstein, Carolyn M. *Do It Yourself: Home Improvement in 20th-Century America*. New York: Princeton Architectural Press, 1998.

Haring, Kristen. "The 'Free Men' of Ham Radio: How a Technical Hobby Provided Social and Spatial Distance." *Technology and Culture*, 44 (October 2003), 734–761.

Haring, Kristen. *Ham Radio's Technical Culture*. Cambridge: MIT Press, 2007.

Harris, Elizabeth. *The Boy and His Press*. Washington, DC: Smithsonian Institution Press, 1992.

Jerome, John. *Truck: On Rebuilding a Worn Out Pickup, and Other Post-Technological Adventures*. Boston: Houghton Mifflin, 1977.

Knightley, Keir. "'Turn it down! She shrieked': Gender, Domestic Space, and High Fidelity." *Popular Music*, 15, no. 2 (1996), 149–177.

Maines, Rachel P. *Hedonizing Technologies: Paths to Pleasure in Hobbies and Leisure*. Baltimore: Johns Hopkins University Press, 2009.

May, Elaine Tyler. *Homeward Bound: American Families in the Cold War Era*. New York: Basic Books, 1988.

Moorhouse, H. F. "The 'Work' Ethic and 'Leisure' Activity: The Hot Rod in Post-War America." In *Historical Meanings of Work*, ed. Patrick Joyce (Cambridge: Cambridge University Press, 1987), pp. 237–257, 306–309.

Silverstein, Ken. *The Radioactive Boy Scout: The True Story of a Boy and His Backyard Reactor*. New York: Random House, 2004.

Games and Sports

Betts, John Rickards. "The Technological Revolution and the Rise of Sport, 1850–1900." *Mississippi Valley Historical Review*, 40 (September 1953), 231–256.

Hardy, Stephen. "'Adopted By All the Leading Clubs': Sporting Goods and the Shaping of Leisure, 1800–1900." In *For Fun and Profit: The Transformation of Leisure into Consumption*, ed. Richard Butsch (Philadelphia: Temple University Press, 1990), pp. 71–101.

Hinden, Rick. "Take Me Back to the Ball Park." *Historic Preservation*, 31 (July–August 1979), 42–50.

Margolies, John. *John Margolies's Miniature Golf*. New York: Abbeville Press, 1987.

McGee, W J. "Fifty Years of American Science." *Atlantic Monthly*, 82 (September 1898), 307–320.

Nasaw, David. *Children of the City: At Work & At Play*. New York: Anchor Press/Doubleday, 1985.

Ritchie, Andrew. *Major Taylor: The Extraordinary Career of a Champion Bicycle Racer*. Baltimore: Johns Hopkins University Press, 1996.

Shergold, Peter R. "The Growth of American Spectator Sport: A Technological Perspective." In *Sport in History: The Making of Modern Sporting History*, ed. Richard Cashman and Michael McKernan (St. Lucia: University of Queensland Press, 1979), pp. 21–42.

Surdam, David G. "Television and Minor League Baseball: Changing Patterns of Leisure in Postwar America." *Journal of Sports Economics*, 6 (February 2005), 61–77.

Willard, Frances E. *How I Learned to Ride the Bicycle: Reflections of an Influential Woman*. Introduction by Edith Mayo, ed. Caro O'Hare. Sunnyvale: Fair Oaks Publishing, 1991.

Extreme and (Sometimes) Impolite Sports

Bates, Carl. "How to Build a Glider." *The Boy Mechanic: 700 Things for Boys to Do*. Chicago: Popular Mechanics Press, 1913.

Borden, Iain. *Skateboarding, Space and the City: Architecture and the Body*. Oxford: Berg, 2001.

Finney, Ben and James D. Houston. *Surfing: A History of the Ancient Hawaiian Sport*. Rev. ed. San Francisco: Pomegranate Artbooks, 1996.

Laviolette, Patrick. *Extreme Landscapes of Leisure: Not a Hap-Hazardous Sport*. Farnham: Ashgate Publishing, 2011.

Stranger, Mark. *Surfing Life: Surface, Substructure and the Commodification of the Sublime*. Farnham: Ashgate Publishing, 2011.

Westwick, Peter and Peter Neushul. *The World in the Curl: An Unconventional History of Surfing*. New York: Random House, 2013.

Electronic Games

Cassell, Justine and Henry Jenkins, eds. *From Barbie to Mortal Kombat: Gender and Computer Games*. Cambridge: MIT Press, 1998.

Chidambaram, Laku and Ilze Zigurs. *Our Virtual World: The Transformation of Work, Play and Life via Technology*. Hershey: Idea Group Publishing, 2001.

Fey, Marshall A. "Charles Fey and San Francisco's Liberty Bell Slot Machine." *California Historical Quarterly*, 54 (Spring 1975), 57–62.

Huntemann, Nina B. and Matthew Thomas Payne, eds. *Joystick Soldiers: The Politics of Play in Military Video Games*. New York: Routledge, 2009.

Kerr, Aphra. *The Business and Culture of Digital Games: Gamework/Gameplay*. London: Sage Publications, 2006.

Levy, Steven. *Hackers: Heroes of the Computer Revolution*. New York: Anchor Press/Doubleday, 1984.

Mahan, Jeffrey B. "Federal Copyright Law in the Computer Era: Protection for the Authors of Video Games." *University of Puget Sound Law Review*, 425 (1983–84), 425–440.

Newman, James. *Videogames*. London: Routledge, 2004.

Poole, Steven. *Trigger Happy: Videogames and the Entertainment Revolution*. New York: Arcade Publishing, 2000.

Schull, Natasha Dow. "Digital Gambling: The Coincidence of Desire and Design." *Annals of the AAPSS*, 597 (January 2005), 65–81.

Schwartz, Frederic D. "The Patriarch of Pong." *American Heritage Invention and Technology*, 6 (Fall 1990), 64.

Wolf, Mark J. P. *The Medium of the Video Game.* Austin: University of Texas Press, 2001.

Index